SMALL SCALE BREEDING

HORSEKEEPING

SMALL SCALE BREEDING

Ray Saunders

STERLING PUBLISHING CO., INC. NEW YORK

Other titles in the **Horsekeeping** *series*
Ownership, Stabling and Feeding
Management: Ailments and Injuries
Riding and Training

Published in the United States and Canada by
Sterling Publishing Co Inc
2 Park Avenue, New York, New York 10016

Produced for the publisher by The Baton Press
44 Holden Park Road, Southborough
Tunbridge Wells, Kent TN4 0ER

Copyright © Ray Saunders 1985

All rights reserved. No part of this publication may be reproduced, stored in a retrieval system, or transmitted, in any form or by any means, electronic, mechanical, photocopying, recording or otherwise, without the prior consent of The Baton Press.

ISBN 0-8069-7680-2 (Softcover)

Printed and bound in Great Britain by
The Garden City Press Limited
Letchworth, Hertfordshire SG6 1JS

Contents

		Page
1	Reflections about breeding	9
2	Land buildings	17
3	Sending your mare to stud	33
4	Foaling	43
5	Raising young stock	58
6	Weaning and afterwards	69
7	Feeding	81
8	Stallions	88
9	Ailments	99
	Index	107

(Colour plate section between pages 64 and 65)

Acknowledgements

My thanks to Mr. George Carter B.V.Sc., M.R.C.V.S., and his colleague Mr. Philip Browne B.V.Sc., M.R.C.V.S., for their help in allowing me to photograph the operations shown and for their advice and information, and my thanks also to my editor, Candida Hunt, for checking my manuscript and arranging the layout for the book. Jack Neal once again helped me to produce all the photographs, for which I am grateful, and my wife Jill was invaluable in all departments of the work involved, both with the horse management and the photographs.

1 Reflections about breeding

Before embarking on a breeding programme it is important to reflect on whether you really want to take the time, trouble and expense of breeding your own foal or whether it would be more prudent to purchase a youngster and train it for your intended purpose. It is easy to let your heart rule your head if you own a mare and out of sentiment feel you would like to have a foal from her, but this is not the best approach to breeding. It sounds nice to hear people say 'I bred it myself', and to see a mare and newly born foal together excites a feeling of joy and envy in everyone interested in horses, but in reality the breeding of horses is a demanding, time-consuming, costly and often frustrating business. Why, then, breed? It certainly will not be in order to make a profit, for unless you are breeding really top class Thoroughbreds (and that end of the breeding scale will be far out of reach of those likely to read this book) or can produce top quality animals of a breed currently fashionable, you will not show a profit from small-scale breeding. If you already own a mare and have the necessary facilities to breed a foal, the stud fee, feeding, veterinary charges and the labour involved all still have to be considered over a long period. When this is added up over the five years that it takes from when you first send your mare to stud until the foal she produces is old enough to begin its proper training, you will realize that to sell this offspring a considerable price would be needed to recoup your outlay.

Another thing to remember is that it costs as much in time and money to produce inferior stock as it does to produce something of reasonable quality, so breeding from 'weeds' is to be discouraged if only because it makes practical financial sense not to do so. Having said that, I do not think one can entirely rule out the sentimental aspect of breeding, nor in my view should one do so, provided that other factors are taken into account and the potential breeder ensures that all the facilities needed are at hand. The type of animal it is hoped to produce and the tasks the offspring will be destined to

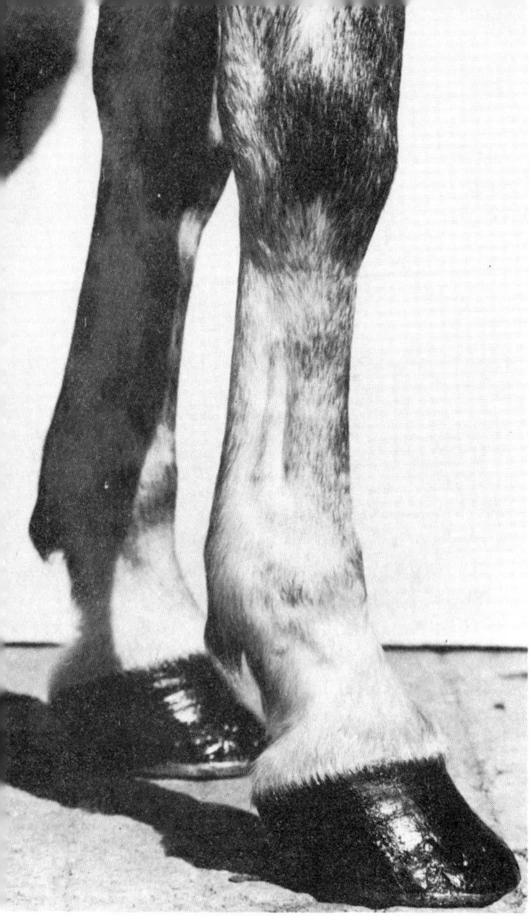

Lateral view of forelegs showing short cannons with a good span of bone and strong tendons. The knees are large and flat, and there is no suspicion of the tendons being 'tied in' below the knee.

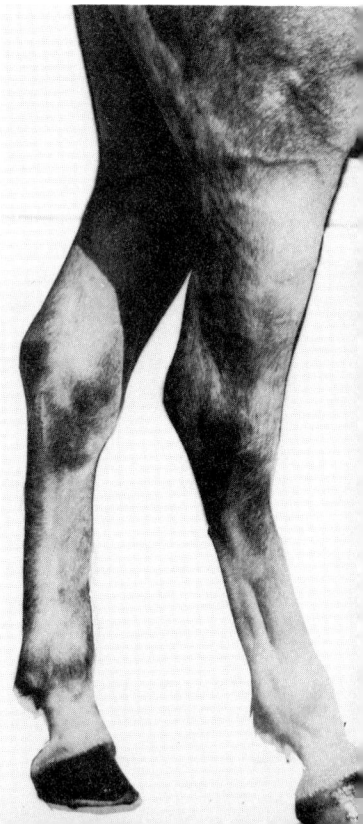

Lateral view showing good, well-matched hocks. Note the well-defined points and overall strength coupled with strong, clean cannons and tendons.

undertake must also be clearly established in the breeder's mind. With that decided upon, and with the right mare and choice of stallion there is no reason why you should not go ahead and successfully breed a foal of your own. The purpose of this book is to help you do just that.

The right type and characteristics

Be realistic about the type of animal you require. Size is obviously one consideration, and although not guaranteed by the choice of parents the offspring will often come closer to that of the dam than the sire. Stallions are selected to try and improve upon the height of certain breeds, but it is not advisable to mate a very large stallion to a very small mare in the hope of producing a large offspring. Fortunately nature usually intervenes if this is done and regulates the size of the foal in accordance with the mare's pelvic region, through which the foal must pass in natural birth; however, complications can result when there are mismatchings with extreme size differences. Small stallions put to large mares (provided that the stallion is tall enough to mount and penetrate the mare successfully) will often result in a large foal – as, for example, with the popular Anglo-Arab crosses. Small Arab stallions put to large Thoroughbred mares will often produce progeny more up to the TB size when they mature. Not only the size of the actual parents enters into the equation, however, as the height of the line of ancestors on both sides must be considered. As with other characteristics, we are better able to judge the probable outcome when breeding from stock of registered parentage because we know with certainty the details of the parent's forebears.

Apart from height, the weight-carrying ability of the type you require will also have to be considered when selecting your mating partners. It is no good finishing up with something that has long legs lacking 'bone', or a large-bodied animal on legs not in keeping with its robustness. A measurement is taken round the cannon bone of the foreleg just below the knee and according to the height and general size and conformation of the animal this measurement is used to judge its weight-carrying capability. For example, a good hunter type that is up to weight would be expected to have 9 inches (22.8 cm) of 'bone', that is to say the circumference of the foreleg just below the knee measures 9 in. This is a general indication that the particular horse has the strength of frame necessary to carry substantial weight throughout the demanding exertions of the hunting field without detriment to its well-being. Of course there are other considerations, such as fitness and soundness of wind, but the first thing looked for in a horse 'up to weight' is this reference to its bone. Size alone in this area is not the end of it, however, as the quality of the bone is important and this will be determined by the tendencies of its parents and proper feeding when young. Another very important aspect is the shape of the leg just below the knee where the measurement is taken. If the size there is much less than the size lower down the cannon bone it represents the condition known as 'tied in below the knee'. This is not good conformation, as it indicates not only a lack of bone size but also suggests weak tendons which, when put to the test of demanding work, will be likely to break down. When selecting for breeding we must look to this part and consider it in the overall size and substance of each partner and try to select both mare and stallion that are well endowed or, as next best, to select one mate that is exceptionally good in this respect to make up for the other if it is not. You can then hope to breed something that comes closer to the conformation of the better parent. Selecting one breeding partner that is strong on certain points does in some cases 'breed out' the weakness shown by the other partner, especially when it is the stallion that possesses the strong feature. It is important to avoid breeding from two animals that are both severely lacking in a particular characteristic; if both parents lack bone, for example, it is a safe bet that so will their offspring. In theory, this should enable more perfectly developed stock to be produced and is the basis of line-breeding, in which certain pronounced characteristics in given animals are matched with those in others so that they are developed and made even stronger with each successive generation. Speed and stamina are two examples that come to mind in the breeding of racehorses, though I do not intend to deal with racehorse breeding as such, as it is a specialized field needing deep analysis beyond the scope of this book. Nevertheless, even the small-scale breeder will do well to consider carefully the history of both the mare and the stallion that it is hoped to breed from. This will give a better idea of what is likely to result from a particular crossing, and eliminate as far as possible the possibility of something entirely unsuitable being produced.

Selection

Try to develop an eye for the type of horses or ponies that interest you, and look around at the various shows, events and sales in your area to help you acquire the skill and knowledge needed to make a sound judgement. Much can be learned about the types and breeds that will suit your purpose. You may want to find a suitable mare from which to breed, and sales will give you the chance to compare a whole range of individuals as well as talking to the vendors and inspecting each animal at first hand. It might also give you the chance to select a suitable animal at the right price. Any defects that are present can then be assessed to ensure they are of a nature that will not constitute a serious fault for your purpose. It may be that a particular mare is for sale because she has a soundness problem as the result of an injury, but that this would not interfere with her usefulness as a brood mare. As long as the injury was not caused by faulty conformation, such a mare can make a very good and inexpensive dam, as damage brought about through injury cannot be passed on to her offspring.

If you already own a mare from which you are thinking of breeding, be honest with yourself and recognize her defects as well as her good qualities, and then set about 'matching out' the bad points when looking for a suitable stallion to mate her with. The foal will draw its characteristics from the blood of both its parents and thus from the heredity of both. The better bred each parent the more truly fixed will be its characteristics, and those of the better bred parent will usually predominate in the foal. The genes carried in each parent are responsible for fixing what type of animal the offspring will be, but these can jump one or even several generations, occasionally producing what is known as a throwback, which may be a common or ugly foal that bears no resemblance to its parents. This will naturally cause great disappointment; it is one of the hazards and frustrations of breeding. Much less likely is the possibility of producing a throwback of unusual quality from some outstanding individual in the line whose virtues were hidden in the foal's parents but were carried in their genes.

It is true to say that the more precautions we take in selecting breeding partners the more the law of averages will work in our favour to produce the desired result. Breeding from registered stock that carries the desirable characteristics from carefully selected breeding over many generations gives the best chance of ascertaining in advance the type of foal we can expect to produce. The lesson is to select at least one parent (usually the stallion) that carries impeccable ancestry and to match the parents so that the good points of the better bred one predominate over the points you are seeking to improve upon in the other. Likewise, try to match the sire and dam in overall compatibility. If your mare has a long back (nearly always longer in mares in order to accommodate their offspring), or if her head is ugly, her quarters weak, she is lacking in bone or has a 'marish' disposition with a tendency to be temperamental or to kick, but you really *must* breed from her, be sure to select a stallion that is impeccable in those points backed up by an ancestry of similarly good characteristics.

Conformation

The desirability of good conformation when mating two individuals is important not only to produce stock of good proportions, weight-carrying potential and soundness but also needs to be considered when wanting to produce a type of animal suitable for a particuar objective. One would obviously need to have near perfect conformation in animals destined for the show ring; jumping will require animals to have wide, strong quarters; gymnastic high school work requires powerful hocks and haunches; eventing or cross-country pursuits require speed and deep chests and well-sprung ribs for stamina. Conformation also determines a horse's action, its freedom of movement, suppleness and comfort under saddle, and thus its ability to be trained successfully for its chosen purpose. Although other factors come into it, without the right conformation no amount of training will produce significant improvement in the horse's ability. Equally important, and even more difficult to produce, is the willingness to perform, perhaps best summed up by the word 'verve'. I hold firm the view that champions are born and not made, and although training is necessary – indeed vital – to produce and refine the inherent potential, without inborn ability or verve no creature, animal or human, will excel at any sport, no matter how good or prolonged its training.

If you wish to breed a foal with a given objective in mind you must first acquaint yourself with the necessary knowledge about the subject of conformation generally. It cannot be learned just by reading a book; this is where visiting shows can pay dividends. In order to help the reader understand the points to be looked for I will explain the various parts of the horse and the importance of their specific functions. I have already dealt with bone and the desirability of having size and strength in the area of the foreleg below the knee, especially in animals that are intended for use in rough and demanding riding in conditions that are likely to play havoc with its legs. All I need to add is that the length of the lower foreleg between the knee and the fetlock joint representing the cannon bone should be short in relation to the upper leg, and that it should be firmly defined with flat sides rather than round and fleshy. This gives the desired strength of bone and tendons so necessary in preventing lameness as demands upon the horse are increased.

The hocks also warrant special attention, as injury and disease in this area can be the result of poor conformation. Ideally both parents should have strong, well formed hocks, but if your mare is lacking in this region – perhaps having slightly overbent hocks (sickle hocks) or carrying the points close together (cow hocks) – then choose a stallion possessing exceptionally good hocks. If the mare has a over-long back with a tendency to a sway back, then look for a close-coupled, well made-up stallion to put her to.

Likewise, all other parts of any mare being considered as a

brood mare should be scrutinized and where defects are present these should be matched out by the stallion. It is possible to be too critical of a mare, especially if each point of conformation is looked at separately and not as part of the whole. Watch the horse in action and form an overall opinion of it; if it moves well and on standing presents a picture of symmetry and balance, small defects of conformation are probably of no great significance. Only if you are very lucky will you find a potential brood mare that is near perfect in every way; if you have one that is fairly pleasing overall with no serious defects it should be possible to match out her imperfections against the chosen mating partner.

Much depends on the purpose we are breeding for. Speed, courage and stamina will be needed for competition animals, whereas these will be minor considerations in the general riding horse or pony; excellent temperament and kind disposition will represent major priorities in the latter while they might take second place when breeding eventers. Stallions will of course always be more excitable and show more mettle than other horses, but those of very high spirits would not be a good choice unless one particularly wanted his other qualities in a high-spirited offspring; a 'scatty' mare is also not good as a brood mare as the offspring will often follow its dam in this characteristic. Kind, placid mares can, even when put to a stallion of rather excitable nature, often produce an equable offspring.

Those interested in producing a dressage prospect will need a foal not only with an inherited calm temperament and poise but also one that will grow up with as near faultless action as possible. Straight movement will be of prime importance, and this should be looked for in both mare and stallion. If the mare is otherwise suitable but has a slight 'dishing' to her action, choosing a stallion that moves very straight and true may produce a foal that inherits this quality from his sire.

The order of preference for the qualities required in any horse or pony will differ among the many people destined to be small-scale breeders. The highest on my list is temperament, for no matter how good a horse is in other respects if it has an unsuitable temperament it cannot be properly employed and all the pleasure in riding and training it will be taken away. With horses that one intends to breed from, either a mare to be purchased or the stallion one considers using, the temperament of the animal will have to be judged beforehand. I personally like to see a good 'kind' eye that is large and steady in its gaze. Small, piggy eyes often indicate meanness; eyes that are rolled, showing much of the whites, can indicate an unpleasant nature; a vague, distant look from the animal will often mean it has psychological disturbances with a failure to concentrate.

Next in the order of qualities I put soundness, except where I can be sure that any unsoundness is the result of a purely accidental injury. It may be that some forms of unsoundness are less important or may not materialize at all in animals not required for the more demanding pursuits, but nevertheless a fit and healthy animal adds much to the pleasure of ownership. Soundness of constitution as well as conformation is equally important, and good 'doers' are much more enjoyable to own.

Stamina, performance, courage and action all follow the first two requisites, depending for their importance on the type of animal being bred and the goals one is aiming for in its future training.

2 Land and buildings

Before breeding can be attempted even on a very small scale it is necessary to own or to have the use of sufficient land and buildings to accommodate the mare and foal. For the breeding of horses one should ideally have large pastures and a good selection of buildings such as loose boxes, barns and shelters, but many small-scale breeders manage quite well with more limited facilities. My own establishment only comprises some 10 acres (4 hectares), but I do see to it that I am not overstocked, and I have also carried out a continuous programme of improvement to ensure that each facility, land or building, is properly maintained and functions in the best interests of both horses and handlers. I will explain in detail what needs to be done and what facilities are required, but generally speaking the two things to be avoided are over-crowding and mud – and the former always contributes to the latter.

First the land. In my opinion it is unwise to start breeding unless you have at least 2 acres (0.8 ha) of good pasture to accommodate each mare and foal. Some will doubtless claim they have managed with less, but I do not believe that less than this is practical or wise. In the first place it must be remembered that every foal will grow into a mature animal, so if you intend to keep the mare you will soon have two fully grown animals. Second, the foal must have room to grow and develop and will need separate quarters at weaning time; third, there are other things to consider such as resting grassland to keep it in good heart. The more grassland you have per animal the better, as with ample grazing you can divide and alternate the use of paddocks, so preventing them from becoming coarse and horse-sick. There is no doubt that youngsters do best when they have room to move about freely in unconfined areas, where possible with others of their own age. This is not to say that a single mare and foal cannot be successfully managed, but thought and care are needed if the end result is to be a contented, well-developed youngster.

The single most important factor to consider with any grass paddock or pasture is drainage. This is especially true because the best type of paddock is one that is on level land. Level paddocks can also mean that they are low lying and so might be prone to bad drainage. This can lead to much poaching of the ground, which causes the grasses to be killed and only weeds to grow back, and if it has a tendency to be marshy it will contain coarse, tufty grass and reed-like herbages that have little or no feed value, especially for growing foals. The best grass for horses is found over a limestone subsoil, and it is no accident that most of the world's top breeding establishments are found in these areas. A limestone subsoil will mean good drainage, and a high level of lime and calcium will be present in the grasses, providing the horse with the essential elements necessary for the formation of strong bone. Other soils can be satisfactory, especially if managed properly, but if the land becomes waterlogged the goodness will be leached away, producing sourness so the horse will not thrive. Heavy clay can be particularly troublesome, as in wet weather it will become saturated and waterlogged without air, killing off healthy plant life, while in hot, dry weather it will bake like concrete, to leave fissures and large gaps where it cracks that can trap and break a leg.

Drainage, by either ditching or the laying of land drains, can be a very costly operation and in many cases will be beyond the financial resources of the small-scale breeder, but to ignore this fundamental requirement and try to use unsuitable land will cause much heartache and frustration when winter sets in. If it is possible to house the mare and foal in a large covered area or loose box for the worst of the weather this will be helpful, and indeed will be necessary in many areas during the worst months. However, to deny room to exercise freely for too long a period is not advisable, as foals and young stock must be able to romp and run in order to develop. To sum up I will say that if you cannot be sure of the use of a reasonably level paddock containing a good mixture of nutritious grasses that is sufficiently well drained to stand up to a wet winter, and if this cannot be backed up by the use of some covered area during the worst months, then your likelihood of becoming a successful breeder both in terms of the end product and the enjoyment of doing it are minimal.

Steeply sloping fields can be dangerous when high spirited animals gallop about, and will also affect the action of the developing animal as its movements must always be performed over uneven ground. It will also put an unwelcome strain on young joints and ligaments that will probably show itself in leg troubles in later

Paddocks can be divided by fences of three strands of plain wire. Corner posts should be well braced and wire tensioners used to keep each strand tight. Take care that no sharp edges or loose wires are left that can cause injury to inquisitive young stock.

life. This will apply less to pony breeds, which are often bred in hilly country, but they mostly have large areas to roam over with some level land and not sloping areas alone.

Your pasture will need to be divided into at least two separate paddocks, so that when one part is grazed down it can be rested for the grass to recover while the other paddock is being used. This will prevent the horses from eating favoured places bare and encourage more even grazing. Horses are notoriously bad grazers, in that they will select special parts of any paddock that they find especially palatable and then continuously graze on these relatively small areas, eating the grass right down to bare earth. The rest of the pasture, even though it looks equally good, will be left to grow lank, and one is then left with the problem of several small areas of bare ground and larger areas of coarse growth. Smaller enclosures of, say, 1¼ acres (0.5 ha) will also make the catching up of horses and young stock much easier, but will still be large enough for them to run loose at exercise. Obviously the more horses that you have together the more room they will need, but the size quoted will be just about large enough for up to four animals – two mares and two foals – at a time. When moving them onto another paddock, say after two or three weeks' use depending on the season, work will be needed to bring it back to good condition. First you will need to harrow it to break up and scatter the piles of droppings so that air and light can get to the grasses beneath; failure to do this will cause them to yellow and die off, taking a long time to recover and then often coming back much coarser than before. Harrowing will also expose worm larvae to the sun and birds which will help in the control and prevention of worm build-up. Another essential part of the pasture's management is to top off all the taller weeds and coarse herbage that are left uneaten, which will encourage the growth of the finer grasses beneath. If the right time is chosen this topping will also prevent the weeds from seeding, and so prevent their spreading to other areas. Thistles and docks, to name just two, can be controlled in this way without the use of poisons, saving both expense and the risk of harmful side-effects to animals and humans. It is surprising how after only two or three seasons a quite badly weed-infested pasture can be brought back to good condition if this work is systematically carried out. I use an old tractor behind which I attach a disc-type mower that I adjust to cut about four to six inches above the ground, and fix a set of old spike harrows (chain harrows are better but more expensive, whereas the obsolete type can be bought for a scrap metal price) behind the mower. Mowing and harrowing can then be carried out as one operation, with considerable saving in time and fuel.

If there has been a period of wet weather, for example a wet spring, the field will also require rolling to replace the grasses and promote growth, and also to smooth out the potholes and prevent these from becoming dangerous when baked by summer sun, when they could cause a broken leg. You will have to choose the right time to do this, a time when the ground is not sodden nor completely dried out, and soils will vary – the timing for clay is especially tricky. Winter attention to grassland is minimal, and you will often do best by leaving it alone; never touch it when it is frozen, or in spring when it is covered by frost.

Depending on your local conditions and type of subsoil there will be a need for you to have the fields limed. Horses and ponies graze a lot of lime in the form of calcium from the pasture, and unless you are on very chalky soil it will be necessary to have this replaced. If you are not sure about your fields as far as this requirement is concerned it is a simple matter to contact a firm of agricultural agents in your area who will carry out field liming by contract. They will test the soil and advise you on the amount of lime required and the price of delivery and spreading. I have fairly neutral soil, and have all my grassland limed at the rate of two tons to the acre once every three years. Very acid or wet land will require it more often. Always choose dry, windless days for this to

Cutting weeds to prevent seeding and topping coarse grass to encourage the finer grasses is carried out with an old tractor and disc-type mower. Harrows attached behind the mower break up and scatter the piles of droppings and aerate the pasture.

A close-up of the rotary mower, showing the adjustable cutting discs and the old type of harrows fitted behind.

be done or half your money's worth will be blown over to your neighbour's fields.

Making your own hay

If you have enough grass to allow a field to be harrowed and rolled and then shut off from late autumn till the following early summer you can make your own hay. Again you can usually arrange for this to be done by an agricultural contractor or local farmer, as it is not worth the capital outlay for haymaking equipment unless you have a large stud and plenty of grassland. Usually you will be charged by the acre for the cutting, then hourly for mechanical tossing and turning with baling done at so much per bale. In an average season the cutting, turning and drying and then baling is completed in a week to ten days, depending on the weather; bales made per acre will work out at between 60 and 100; the overall cost will work out as a considerable saving over buying hay from a farmer or merchant. It also gives great pleasure to make and use one's own hay, especially when you know that it has cost you only between a quarter and a sixth of the price of buying it.

If you intend to make hay in this way, be sure to arrange it early with whoever you get to do the work, as when the haymaking season begins everyone wants it done at once and there is a premium on the machinery. You will not want it cut until the weather is right, when many others in your locality will also be clamouring for the machinery; if you have made prior arrangements you will be much more likely to be at the top of the list. It is always a great relief to see it baled and carried off the field before the weather changes, as rain can spoil it or greatly reduce its feed value once the grass is cut and drying. Made successfully, it is a most important factor in the economics of breeding horses. You will of course need a barn for storage; this can be built using materials and methods that are inexpensive, as I explained in the second book of the 'Horsekeeping' series, *Management, Ailments and Injuries*. There are a few important tips I can give to those making hay for the first time.

1. Try not to have the grass cut until the ground beneath is dry, as the cut grass will not dry where it is in contact with wet earth; sun on the top of it will also cause the underneath to sweat, and produce mildew.
2. On no account bale it until it is dry, and make sure there are no wet patches as this will also cause mildew when the bales are stacked.
3. If it is a very hot year do not leave the grass to dry for so long that it is baked and bleached right out before baling; the mineral and protein content will be lost if this happens, and the feed value will be reduced to that of roughage.

4. Do not have it baled up too tightly as this can compress it too much and prevent it from 'making' evenly with a tendency to produce mould between the compacted segments.
5. Do not have the bales too large as they will be too heavy for you to handle and stack.
6. Keep animals off the field that has been cut for hay until it has recovered again and grown back to four inches or so in length.
7. After haymaking is a good time to lime the field if this is to be done and animals should then be kept off while it is recovering and until there has been a good rain to wash the lime in.

There are those who spend considerable sums on artificial fertilizers for their fields, and if you feel this is necessary you should seek the advice of an agricultural service who can advise you on when and by how much this is required. Personally I do not use any fertizer other than natural additions to the soil produced from droppings. I have several reasons for this. I am not convinced that fertilizers are necessary if a proper programme of harrowing, rolling and resting of the pastures is carried out – the manager of what are considered to be England's finest pastures is on record as saying they have never been fertilized in two hundred years, merely limed and managed in the way I have described; once the balance of the soil is disrupted, as happens when artificial fertilizers are added, it becomes necessary to fertilize continuously in order to correct and make up for imbalances and deficiencies. I also hold the view that the application of artificial fertilizers can in certain circumstances affect the horse's health and well-being, and I subscribe to the widespread belief that such 'mystery' ailments as azoturia can sometimes be attributed to this factor.

Before leaving the subject of grassland management I must mention another useful way of improving the balance and maintaining it in good heart. This is to intergraze cattle if your paddocks are large enough to provide spare capacity for grazing. Only use cattle that have been dehorned, as otherwise there is a danger of playful young horses becoming gored by angry cows. Cattle and horses do not recycle the same parasites, which is helpful in worm control, and will also graze over each other's droppings without being put off as they are over their own. Another advantage of this intergrazing is that cows need longer grass as they crop it off by using their tongue to wrap around the stems, whereas horses use their front teeth to tear off the grass near to the ground. Cows will therefore clear the longer, tufty grasses that are often left by horses.

Provision of water

Another essential item that must be considered is the provision of an ample supply of drinking water. In-foal mares and those suckling their foals drink more frequently and consume more water than at other times, and likewise drink more than other horses. If you are lucky and have clean, fresh water close to your paddocks this supply can often be utilized if it is constant and does not dry up during long dry spells. Rivers and ditches actually running through paddocks can be dangerous, however, and I have known many people to lose mares and young stock that have fallen into these and either drowned or injured themselves struggling to get out. A far safer method is to use the tactic I employed to provide constant fresh water without risk. I ran a length of 1-inch (2.5 cm) alkathene pipe from a dam in a small stream (built of concrete to form a small reservoir) through the fence to my paddock. It was buried about a foot underground and taken to a galvanized tank that was positioned so that it could be reached by horses on either side of another fence that divided two paddocks. This tank stood lower than the dam, and the pipe could be fixed into the bottom of the tank with an overflow pipe taken from the top of the tank and led back to return the overflow water to the stream lower down. This works perfectly, with continuous fresh water running through the tank so that it never becomes stagnant. Running water is also less likely to freeze in winter, and as a further precaution against freezing I have also lagged and boxed in the pipe at the tank.

If no fresh water exists you will either have to transport water to your field – a laborious and time-consuming task – or resort to a mains supply if there is one near enough to tap onto. Water piped from the mains will be somewhat costly depending on your area, as there is usually a standing charge for each outside supply plus a metered charge for the amount of water consumed.

The site chosen for a water trough also needs careful planning as it is a place where stock will naturally congregate, causing a lot of poaching to the surrounding area. For this reason it is advisable to put down a 6-inch (15 cm) dressing of stone or rubble (not rubbish containing nails and glass) that will bed down to a hard standing that will not become waterlogged. I contained the area of stone around my water tank with old railway timbers laid flat to form a rectangle, and then infilled with 3-inch (7.5 cm) stone chippings; failure to do this can result in the stones being scattered and their effectiveness lost.

Fences

The type of fencing used to surround or divide paddocks that are to enclose mares and young stock needs careful consideration, as with these animals there is much greater risk of trouble and possible injury than with adult grazing animals. Posts and rails are probably the safest, but even using the cost-saving methods to build these that I described in the first book of this series,

A water tank positioned so that stock can drink from both sides of the paddock fence. Note the infill of stone, which is contained by a rectangle of old railway timbers, to prevent poaching. The water pipe was later lagged and the tank boxed in to prevent freezing in winter.

A large, strong field shelter with plenty of height and a wide entrance to prevent crowding; storage for hay is also provided (right of picture). The large tubular gate joins on to the paddock dividing fence (far right) and can also be swung across the shelter entrance to shut stock in for examination or during bad weather.

A close-up of the combined manger and hayrack fitted in the field shelter, made from old timber and wire stock fencing.

Ownership, Stabling and Feeding, it will still prove a considerable financial investment. Apart from boundary fences and hedges, which I shall deal with later, I have found posts and wire satisfactory for making separate enclosures. Only plain wire is used for this – never barbed wire, for sooner or later this will cause a nasty accident to your stock. I use a thick plain wire for the top strand with two thinner strands spaced beneath, depending on the height of the fence. There is one disadvantage with this method and that is when foals are small they do lie down a lot and often tend to do this alongside the fence. They then seem to delight in rolling underneath and getting up on the other side! We had one foal that was so adept at doing this, my wife swore he was a limbo dancer! This has happened in my experience rather frequently but is not too disastrous if you are visiting them two or more times a day. If, however, they should do this and be left for a whole day and cannot get to the mare this is obviously serious as they cannot suckle; this separation will also eventually promote panic in both mare and foal who can then quite easily injure themselves in trying to get back to each other. A better type of fencing, which prevents this from happening but is more expensive to erect is the large, square-holed stock fencing of stout wire; if this is used it should be fitted low enough to the ground to prevent anything getting underneath, but kept clear of the ground so that feet cannot get entangled in it.

For boundaries you cannot beat a good stout, thick hedge that when allowed to grow to sufficient height offers protection from wind and rain, and if it contains a large tree or two also provides welcome shade from the hot sun. Hedges need to be 'knocked back' every year to keep them from encroaching into the paddocks and also to keep them compact and dense enough to give the thick mass of undergrowth needed to afford protection from the elements.

Field shelter

The ability of a horse or pony to survive the rigours of winter in reasonable condition will depend on its constitution and breed. In general, the more finely bred it is, the more it will suffer when the weather turns bad, and youngsters, especially foals in their first winter, will be particularly at risk if left out of doors. To be out in winter for several hours a day will do no harm, but if you must keep them out twenty-four hours every day you will need to provide a field shelter for use during the worst periods of the year. Dry cold or even snow and freezing temperatures will do little harm if your animals are well fed and are left to grow thick winter coats. Driving rain and biting winds, however, will cause them to lose condition no matter how well fed they are, if they are left to endure these conditions for lengthy periods.

Some owners use various types of rug to provide their horses with protection during bad weather, and if properly fitted and seen to regularly, these can be helpful in this respect. I have never personally been an advocate of these for a number of reasons. They can in many instances cause the poor animal more discomfort than they allay – if ill-fitting or allowed to become stiff and heavy with dirt and mud, they can rub nasty raw patches on the animal; if the weather changes, suddenly becoming warmer or with a rise in humidity, the horse can sweat and become uncomfortable, and irritations of this kind will cause the animal to rub or roll, when the rug may be hooked up or twisted out of its proper position, which will often lead to further problems; human nature being what it is, even the most dedicated owners will tend to leave the horse for longer than is wise when it has been left nicely rugged and standing quietly, only to find next day that it is in an awful tangle and has probably been like it for some time.

I much prefer to leave the animal unclipped, with a good coat, to feed it well and provide it with a field shelter. Many people will argue that field shelters are not used and that horses will stay out and suffer in the cold and wet, but I have found that if the weather is bad enough horses do seek refuge in them. Of course the shelter must be correctly sited to give protection from the prevailing wind and rain, and no horse will stand under one that is blasted by the wind driving into its opening or one that is uncomfortably draughty. They can be further encouraged to use a shelter by feeding them their grain or concentrates inside, and making sure that it is always supplied with a plentiful amount of hay. Incidentally, there is much less waste when hay is fed in a shelter, where it will not be scattered and trodden on as happens when fed loose outside. If only a field shelter is available to you for use during the foal's first winter, it is best to shut the mare and foal in during the night by fixing a gate or some poles across the entrance. This is because the mare will be impervious to the bad weather on many occasions and will remain outside with the foal, who will always follow its mother, and so suffer the cold. This can also happen when more than one mare and foal are turned out together, as if the toughest one stays out, the others will stay and suffer rather than go in alone. In this case, if several mares and foals are to be shut in together, the shelter must be large enough for them to move around out of each other's way, in case one becomes jealous in the confined space and takes it out on another mare or foal. Groups of youngsters of the same age herded together will usually sort themselves out without serious harm being done – merely a pecking order being established.

Buildings

The number and type of buildings required will depend on the number and type of horses and ponies you intend to breed. It will range from a single loose box or field shelter combining a space for hay and feed that will suffice for breeding a single foal, to a comprehensive layout of various buildings if breeding is to be carried on regularly from a batch of brood mares forming a small stud. Ideally you cannot have too many buildings, and these should include a range of loose boxes to which a good-sized foaling box should be added, and further augmented with outbuildings such as a barn for hay and straw, an implement shed and a field shelter for your principal paddock. However limited your facilities are, and no matter how small your additions have to be, time spent thinking out and planning the layout will be well rewarded. Breeding and growing animals need space to move around, and if badly planned, the entrances and exits to and from buildings and paddocks can often lead to accidents that, with some forethought, could have been prevented.

Some people prefer to let their mares foal outside unaided, and I must say that left to their own devices normal mares will usually carry out the task successfully without any trouble whatever. There are occasions, however, when a mare needs help, and the presence of immediate assistance can save a foal that would otherwise be lost. Mares, too, are at more risk during foaling than at any other time, and the mortality rate among brood mares is the second highest in the animal kingdom. For those owners wishing to take no chances and who want to sit up and be on hand (foaling often takes place in the early hours), then a foaling box will be necessary. To be of use it must be large enough for the mare to move around freely, and a small loose box is not to be recommended as there is a danger of the mare not being able to get down and up again without difficulty. Another danger in cramped quarters is that she may roll or tread on the foal. She may well get too close to the corner when foaling, and the foal will not be given sufficient space to gain free access into the world. The size the foaling box needs to be will depend to some extent on the size of the animal one is breeding from, but even for pony breeds it should not be less than 10 x 12 ft (about 3 x 4 m), and ideally for all but small ponies a box of 12 x 16 ft (some 4 x 5 m) will be required. If you are lucky enough to own an old barn that can be partitioned off, this will be ideal, as enough room will be available to place straw bales around the sides to exclude draughts and provide protection; there will probably also be enough room to keep your feedstuffs and other essential items all under the same roof.

When constructing a foaling box or converting an old building, always remember that horses are strong and awkward creatures,

When constructing a foaling box allow ample height for a plentiful supply of fresh air. This shows a partly completed dual purpose foaling box or two loose boxes; provision has been made in the roof timbers for a partition to be inserted, and angle irons were fixed to the timber for additional strength.

Water bucket holders are fixed to both sides of the centre timber where a partition can be fitted to convert the foaling box into two loose boxes. Note how the taps are arranged to prevent their being chewed by horses. The stone manger (left) is cemented in and there is another in the opposite corner to the right of the picture.

and materials need to be able to withstand the weight of a leaning or rubbing horse. Supports and the thickness of walls and partitions cannot be made too strong, and attention to this point will undoubtedly save you much trouble and anguish later. Another advantage of converting an old barn is that they are usually high, allowing for a plentiful supply of air and good ventilation without draught. Methods and materials for construction and conversions will be the same as for loose boxes generally, and I dealt fully with building your own stables in the first book of the 'Horsekeeping' series – *Ownership, Stabling and Feeding*. There are some special requirements that apply to accommodating mares and foals that need to be kept in mind when building or converting. Remember that the greatest single thing a horse or pony needs for its health and well-being apart from correct feeding is a plentiful supply of fresh air. It is the one thing you can provide them with, knowing it to be of the utmost benefit without it costing you anything at all. Another important point is to have a large door, and in this case something about 4 ft (1.25 m) in width is required. This width is less essential with an ordinary loose box, but with a foaling box you will find there will be times when you lead the mare out and the foal will try to rush through. This can lead to the foal – or perhaps you – being squashed against the side of the opening if it is not large enough. Another fixture needing special care is the manger. You can provide separate mangers for the mare and foal but if, like me, you intend to have them both feeding from the one, it will need to be large enough for this. In any case, special care must be taken to ensure that mangers are fixed very strongly, so that they cannot be dislodged by a feeding animal. Old earthenware sinks of the type that are large and deep and finished with white glaze make ideal mangers when fitted in the corner of a box. They are very heavy, and are often used for this purpose, being raised off the ground and supported on concrete blocks. Because of their weight, it is often assumed that they will remain supported in this fashion, but in fact this practice is very dangerous. There has been a recent case in my area, where a manger of this kind supported on blocks was knocked into by the mare, causing it to fall onto the foal who happened to be lying on the ground near it. The injury that the foal sustained to its spine meant that it had to be destroyed. The lesson is to always make sure that fittings of this kind are securely held in place, and my own method of ensuring this is to build the supporting blocks into the wall and completely cement everything in, to form a strong permanent fixture. My foaling box was photographed during construction so that my methods can be seen. The arrangement I have for water is also worth noting; the taps are partially enclosed by wooden blocks to prevent them from being chewed, but allowing them to be turned on and off easily;

beneath each tap a bucket holder is fixed so that a large bucket can be kept filled and easily removed for cleaning. The buckets and the taps are placed in the centre of the foaling box back wall, where a partition can be installed through the middle, thus converting it into two good-sized loose boxes when not in use for foaling. I designed and built it in this way as a dual-purpose foaling box or two loose boxes, to house a mare and foal each side. With two front doors, the tops of which can be left open, and opening windows along the other side, it is possible to have ample ventilation and a good supply of constantly changing fresh air in hot weather.

Hay racks need to be fitted at a height that can be reached by a foal as it will soon learn to feed from it, and the exercise it gets from copying its mother in pulling out the hay will also develop and strengthen its neck muscles.

Another very important feature is the provision of electric light. If you are to be on hand when needed at foaling time this will be essential, the more so if you are alone with no one to hold a lamp or torch; even for the final inspection each night to see if everything is in order, the presence of an electric light will be appreciated.

Finally, before leaving the subject of buildings let me emphasize the need for forethought and planning. Look to siting before construction and make sure that you are not about to build where flooding will be a problem, and take a good look at the drainage system before converting a building; use solid materials and strong fixings – twice as robust as you first thought necessary. Get all this sorted out before you embark on your breeding programme, and do not leave it to be struggled with when mud and surface water make work difficult, and the situation is further aggravated by additional livestock. Put right any small defects as soon as they are noticed, and do not leave them to degenerate into major repairs or renewals. If you need to construct extra buildings, keep the layout as compact as possible, so you do not waste time moving about from one to the other – this can be especially tiresome in bad weather. At the same time, leave enough room for easy handling without risk of injury to animals or humans, and bear in mind the need for tractors and lorries to move in and out freely.

3 Sending your mare to stud

Many people will want to breed from their own mare by sending her to be serviced by someone else's stallion; for those thinking of owning their own stallion, I will devote a separate chapter later.

Having decided that your mare is of the right type to breed from, you will have to find the right type of stallion to mate her with, and you should make a list of what you consider to be the best choices of stallion in your area. These stallions can then be inspected, and most studs and stallion owners will be pleased to show them off to you – but don't arrive unexpectedly and expect them to drop everything in order for you to view their stock. Always arrange an appointment and then turn up on time, as studs are busy places, and have to organize their day if things are to run smoothly. It is also a good idea to see the stallion's progeny if you can, as some outstanding stallions produce rather indifferent stock, whereas others that are more moderate looking have the ability to produce stock of superior quality. It goes back to the point about genes that I explained in Chapter 1. Any youngsters will of course have drawn their characteristics from the mare as well as the stallion, but if all the young stock show a tendency to have coarse heads and mean eyes, poor hocks or some other pronounced defect, it can be reasonably assumed that the stallion, despite his own good looks and conformation, carries in his genes the propensity to pass on a particular weakness although it is not visible in him.

Another important factor when deciding on a stallion is to look around the stud to check that it is clean and well organized, which will give a good indication of whether or not the owners know their business. If the place is dirty and appears slipshod, it is likely that their approach to the tasks required for successful breeding is also less than dedicated. Sending a mare to stud involves you in time and expense, which will be wasted if the stud does not provide the proper service. In most cases of a mare's failure to breed the fault is with the mare, but even with a fertile mare and stallion, efficient management and the general way that things are run play a great part in your mare being successfully mated. A stud that is well run and takes a pride in its achievements will be anxious to give a good service and maintain the stallion's fertility record.

The proportion of mares that cannot breed or whose fertility is such as to make them difficult breeders is quite high, so do not be surprised if a stud refuses you a nomination if they think your mare falls into this category. This will be more likely to happen with studs offering a 'no foal no fee' service. Maiden mares (those never having visited a stallion) are also an unknown quantity in this respect, and some studs are reluctant to accept them. I have never quite understood why, as they obviously have to go to a stallion for the first time if they are to be mated, and I have not found that they are necessarily more difficult or troublesome to cover. It really boils down to how good the stud management is, and whether they take enough time to give each mare proper attention by not booking in more than can be properly handled. Studs with a popular stallion who know their business and only accept a limited number of mares will probably be booked up well in advance of the mating season, so it is wise to make your arrangements with them by the autumn (fall) of the previous year. Except in the case of Thoroughbreds that are being bred for racing, the normal season for serving the mare will be from spring to midsummer.

Thoroughbreds are all aged one year old on 1 January following their birth, and every 1 January thereafter is their official birthday, making them one year older. In this way a TB foal born, say, in June will only actually be six months old come its first official birthday, when it will be declared one year old. If the animal is required to race as a two-year-old, it will in fact be only twenty-one months old, and competing with others up to six months older – an obvious disadvantage. Breeders of racing stock therefore require their mares to be put in foal as soon after February as possible, so that the foals are born as soon after 1 January the following year as can be managed. Breeders interested in showing will also often try to produce early season foals, so that they are well forward by the time they are entered in the various shows. Having a foal to cope with in the middle of winter is very much more difficult than when the days are longer and warmer, so this particular arrangement is in my opinion best left to the racing and showing fraternity.

For the ordinary horse breeder it is better to have the mare covered much later in the season, so as to produce the foal in the better weather of the following year. The management advantages of this are obvious, and most people will agree that a spring-born foal will have the advantage of long, warm days for its early growth and both it and its dam will benefit from the fresh spring grass. However, there is another arrangement that I believe to be even better than this. It is to have your mare served by the stallion in late summer or early autumn so that she produces the foal at that period the following year. It will depend, of course, on your mare coming into season at this later time of year, but most of them do go on throughout the summer, and in fact will have very strong periods of heat in the late summer – what I call blackberry heat, as it is during the time of the blackberry harvest. I have found this to be particularly true with mares possessing Arab blood, who may show no signs at all in spring and early summer and then make their condition embarrassingly obvious around September. Putting a mare in foal in August or even early September will mean that she will be through the winter with plenty of time on spring and early summer grass to build up her strength, gaining the benefits of this and the sun in the later stages of her pregnancy. This will help to ensure not only that she produces a strong foal, but also has a good supply of milk. If the foal is born in, say, mid-July, it will have the rest of the summer and the autumn in which to develop and grow, and will be almost six months old before the onset of winter. The foal will also benefit from the much warmer and drier weather of the summer/autumn period. Early foals, even if not born until the spring, can suffer from the often very wet and cold spring weather in the U.K. and get a bad start, especially if the mare is foaled outside. Some will argue that flies are more troublesome to the mare and foal at the later time of year, but given good management I have never found this a problem. After all, the animals have to endure the insects at some time whenever they are born and if subjected to them when very young they seem less worried about it than when older.

Another consideration is the age your mare should be when put into foal for the first time. Unlike humans, a mare will be capable of breeding throughout her life, right into old age. I personally believe that the best foals are produced from young mares, but it is also true that many old mares have first foals that turn out perfectly satisfactory. I will say, though, that putting an old mare into foal for the first time means that special care and attention should be given to make sure that she copes without detriment to herself or the offspring.

As to breeding from young mares, there are those that maintain this should not be done until the mare is four or even five years old. The reason for this is said to be because the mare is herself still growing up to that age, and the burden of producing an offspring can stunt her growth and development. I do not believe that this is the case, provided of course that her special needs are recognized, and she is properly fed and looked after. A very young mare can, however, suffer from the effects of producing milk and feeding a late-born foal during the later part of the year, and go back in condition throughout the foal's first winter. It can then be difficult to judge her feeding requirements, as her body will be having to recover from producing a foal as well as developing her natural growth. Certainly I do not think it is wise to put a very young mare straight back into foal again without giving her a 'free' year to catch up. Other than that, I believe that if a filly or a young mare of two or three years of age is put into foal, no harm will be done to her own development with proper care, and any check that she does suffer will be made up later.

When deciding whether or not to put a two- or three-year-old mare into foal one will need to consider her general development, and if she is large enough in the pelvic region to accommodate a growing foal, and present it the following year without difficulty. Your vet can be consulted on this, and he can also carry out an examination to ensure that the mare's internal organs are satisfactorily developed. A stallion that is sensible and not too large for the mare should be chosen so that she will not be frightened and put off future breeding by wild and aggressive mounting at service. This is where a proficient stud will prove helpful, and although some studs hold to the view that three-year-old mares are particularly difficult to get into foal, I have never found this to be a problem when mare and stallion are properly handled, and the right time is chosen for the service.

The cost

Having decided on all this, you will need to know what fee you will be required to pay and what is included in the charge. Studs vary a great deal in their methods of determining payment, and of course the more sought after the stallion the more his service will cost. Some studs advertise the fee with the initials NFNF – 'no foal, no fee', which is self-explanatory and indicates that the stud has faith in their ability to get your mare into foal. However, their fee will probably be high to cover the guarantee. NFFR is another well advertised arrangement, which means 'no foal, free return'. With this arrangement you must clearly determine beforehand whether it is intended that the mare should return to the stallion the same season if she is found to be barren. Some studs will charge part of the fee at the time of service, with the remainder being due on 1 October unless the mare is certified not in foal; another system is to make payment due on the birth of a live foal. My own method is to make a charge payable at service, and this includes one month's keep at grass. If the mare is subsequently tested and found not in foal by 1 October part of the fee is returned to the mare's owner, or I will accept the mare again for service for only the cost of her keep if there is sufficient time that season. Whatever method is adopted by the stud of your choice, be sure that you fully understand the contract you are entering into so that no misunderstanding complicates settlement when the time comes. Another thing you have to decide is whether you intend to return the mare to the stud the following year for the actual foaling and possibly her being put into foal again. Some mare owners like the foaling to take place at the stud and then have the mare and foal home when the foal is some twelve to fourteen days old.

Preparation

The preparation of your mare before going for service is important as it contributes to the success of the operation. The stud will require that your mare is given a veterinary examination and tested to ensure that she is free from disease. Your vet will need to be contacted two months or more before the mare goes to stud so that he can arrange for these tests to be carried out. In the UK there are two requirements: a bacteriological examination of the mare's genital organs, and standard testing for contagious equine metritis and other venereal diseases. It is commonly known as swab testing, and you will be given a certificate for each of the tests; these should be sent or shown to the stud. Where these certify that your mare is clear the stud will accept her without any qualms.

Another part of the mare's preparation is to get her fit. I do not mean by this that she should be racing fit but that she should have been well enough exercised so as not to be grossly overweight. Fat mares are notoriously difficult to get into foal, so make sure she is in trim shape when she goes to the stud. Pony mares that have been turned out on lush spring grass are the most likely offenders, and they should be restricted by keeping them in for part of the day or put onto sparse pasture. It is not a good thing in any case to let them graze uninterruptedly at this time of year for it will make them prone to laminitis. I had a case of an unfortunate owner with a greedy mare that would not stop eating after she was tested in foal and aborted the pregnancy by contracting laminitis. (This mare returned to my stallion the same season, was put in foal again and by being carefully fed produced a beautiful colt foal the following year.)

Most studs also ask you to worm your mare ten days or so before sending her, as this helps to reduce the spread and build-up of worms on their pastures. Another helpful thing is to send her in a good, strong headcollar, and you should make sure this is clearly labelled with her name as this will avoid any risk of her not being correctly identified at the time of service. A lead rope, tail bandage and items such as grooming equipment can also be sent with the mare; these should be placed in a small holdall and clearly labelled with your mare's name. Finally, if your mare has been shod there is the question of whether the shoes are to be removed from all or just her hind feet.

Brood mares

After your mare has visited the stallion she becomes a brood mare. When she arrives home she will probably have 'held' to the service

Before going to service a mare should be swab tested to ensure that she is free from disease. In the UK two certificates, as shown here, are required.

```
                    LABORATORY    REPORT
        BACTERIOLOGICAL EXAMINATION OF EQUINE GENITAL SWABS

                                    Ratley Lodge Laboratory
                                    Ratley
                                    Banbury, Oxon.
                            'Phone Edgehill 501 (4 lines)

    MARE/STALLION'S NAME    Samantha
    OWNER   Blair                    LAB.REF.No. 886·82
    ADDRESS  Shangri La
             Romsley

    STUD
    ADDRESS


    The following swabs   clitoral, cervical, urethral
    were submitted by   R H C Thursley Pelham MRCVS
    on    1·4·82          for aerobic bacterial examination
    RESULT
    Sensitivity:        CEM and other venereal
    Laboratory Comments:  pathogens not isolated.
    Smear-negative and contaminants only
    isolated from cervical and urethral
    swabs.
                                Signed:  Sally K...

    NOTE:  This is NOT a certificate for Contagious Equine Metritis.
           CEM report to follow/enclosed.
```

> **HORSERACE BETTING LEVY BOARD**
>
> **CONTAGIOUS EQUINE METRITIS AND OTHER VENEREAL PATHOGENS**
>
> Standard Certificate
>
> *For use only by Designated Laboratories* December 1981 - November 1982*
>
> The following swabs............*clitoral*............
> contained in transport medium from the mare/~~stallion~~ *Samantha*
> were submitted by......*RHCThursby-Pelham Mews*......for
> bacteriological examination on......*1 April 1982*......(date/s).
>
> **ROBIN HENRY CRESSETT THURSBY-PELHAM**
> I,.. of *Ratley Lodge*
> Laboratories certify that the above swabs were culturally examined for the Contagious Equine Metritis Organism and other venereal pathogens with negative results.
>
> Signed......*RHCThursby-Pelham*......
>
> Date......*6 April 82*......
>
> *A Designated Laboratory is one whose name is published by the Horserace Betting Levy Board in the Veterinary Record on 28th November 1981.

at the twenty-one and forty-two day stages. That is to say she has not come into heat again on what would have been her usual cycle, which is a good indication that she is in foal. Sometimes a mare will 'slip' and come into season again sixty-three days after the service, having held till then, but this is not common. Peculiar things do sometimes happen; I had one mare that came into heat very clearly twenty-one days after the service, so I assumed she had not taken and duly covered her with the stallion again, whereupon she held. I plotted her time to foaling, and to my surprise she then produced her foal the following year bang on time to the first service!

Naturally one wants to be as sure as possible that a mare is in foal, as otherwise a lot of time is wasted. At about six months signs of pregnancy often become apparent, such as the mare's belly becoming wider and dropping lower. This should not be confused with the fat belly of an over-fed mare, which will be uniformly round, whereas when in foal the belly appears more pointed and will often be seen to project out more on one side than the other. From about seven or eight months up to foaling it is often possible to feel and even to see signs of life inside. By placing your hand on the mare's side you can sometimes feel the foal move, and by placing your ear against her side you can at certain times detect quite definite movement, though there are also digestive movements and noises that can be confused with those of a foal. Some mares show no signs whatever, and manage to hide their pregnant condition right up to foaling – it is by no means uncommon for an unsuspecting owner of a recently purchased mare to go out one morning and find her proudly standing beside her newborn foal.

There are veterinary tests that can be carried out to ascertain the mare's condition and pronounce her either in foal or barren. Even these, however, are not infallible, and I have known of mares tested and said to be in foal who subsequently were not – and vice versa. At 42 days after service your vet can give a manual examination, inserting his hand and arm into the mare's rectum to examine the uterus for signs of pregnancy. Between 45 and 90 days a pregnancy diagnosis by blood test can be carried out, and from 100 to 140 days a urine test can be done. There is also an ultrasonic probe that can be given between 18 and 22 days after service that is said to be of great value in saving time at studs where very high-class animals are being produced; I have never personally used this method and I believe it is very expensive.

Whichever pregnancy diagnosis is used the owner should always be prepared for the unexpected, though it is true to say that in most cases the mare will perform as predicted. What I said about not overfeeding the mare before she is sent to stud also applies after she has been put in foal; moderate work up to the fifth month will do her no harm and help to keep her fit; after that she will usually become sluggish and not be a pleasurable ride as she should not be forced into strenuous exercise. Mares that are turned out to grass will keep themselves fit and healthy without needing special attention except to make sure – especially with some pony breeds – that overeating lush grass does not lead to the troubles I have already described. It is a mistake to think that a mare needs to be fed for two in order to produce a strong, healthy foal. Nature plays its part here by arranging the metabolism of the in-foal mare to act more efficiently and derive more nourishment from the normal rations. This will mean that her normal diet will be sufficient for both her and the developing foal. As autumn leads towards winter a field shelter will be needed and supplementary feeding should be started. I like to bring my brood mares in at night at the onset of bad weather and feed them liberal hay rations and a small concentrates feed of approximately 3lb (1.36 kg) – I use half each of bruised oats and crushed barley, mixed together with a handful of molassine meal to which I add a tablespoon of cod liver oil. As the weather worsens I give them another small feed of similar proportions each morning an hour or two before turning them out. The appropriate amounts will of course vary depending on the individual. Some brood mares, depending on

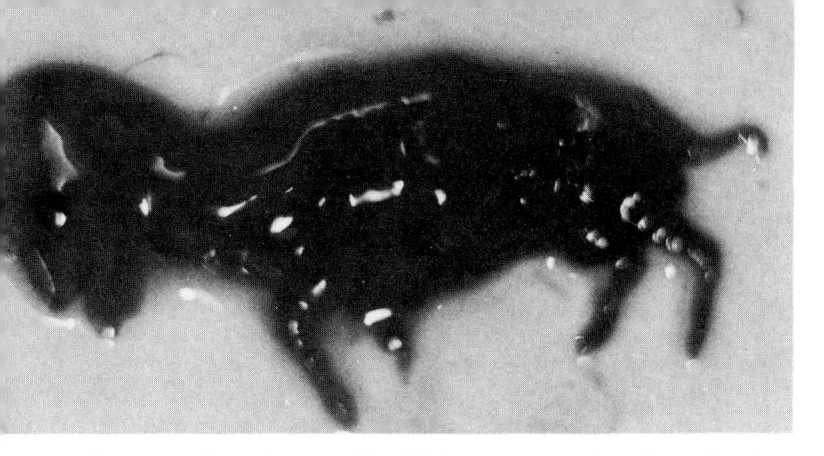

A foetus at about four weeks old from a mare that aborted. These are not often seen as they are usually absorbed or lost in the emitted waste material. Even at this early stage the form of the horse's head, legs and tail can be clearly identified (actual size approx. 40 cm).

their age and size, will need more, and it is particularly important for young brood mares of two or three years of age to be kept in condition as they themselves are still growing and need extra feed to continue their own growth. I should issue a warning here about the feeding blocks that are sometimes used for horses and ponies wintering out. These are compressed forms of 'complete' food that the animal eats by helping itself. There is a similar type that is fed to cattle, but this contains monensin sodium, which is fatal to horses and ponies even in small doses. Another recent cause for concern involves the practice of feeding horses silage, as in some cases it has been found to have serious consequences due to the fermenting process.

One must learn to recognize the right sort of healthy condition, looking for the signs of good health such as a clear, bright eye with pricked ears and a healthy appetite. Movement should be unsluggish and the coat, although it may be shaggy and muddy, still has some 'life' in it. Signs of the mare being unwell will be dullness, sluggish movement, dull eyes with hollows above, a staring coat and a general appearance of dejectedness.

To keep her system clear and assist bowel function I like to feed a bran mash once a week, and this warm meal is always greatly appreciated and eaten with relish. If your mare suddenly leaves her grain ration for no apparent reason make sure that nothing of a nature to taint the feed has inadvertently got into it, as horses can be fussy feeders and be put off by offending ingredients or strange odours. The other necessity is a plentiful supply of fresh, clean drinking water in both the daytime paddock and the night quarters.

When out at grass brood mares will generally do better in company than alone. The company should, however, be of the right type and not given to galloping about. Geldings will not do as they will make a nuisance of themselves and trouble the mare. Even those in an adjoining field can, if allowed to approach within nose-touching distance, cause the mare to be teased sufficiently to slip her foal and come into season. Other brood mares of aimiable temperament are the best companions, but failing that a cow, goat or even sheep will provide some acceptable company. Any cows in the same field should be of the dehorned type for the sake of safety. If you have to keep her on her own, never confine her to a lonely boring existence away from view of all forms of animal life and general activity. I firmly believe that a well managed brood mare that is kept happy and contented is far more likely to produce a worthwhile foal. Not only do I think this to be true of the physical attributes, which will obviously benefit from good feeding, but also that the foal's temperament is far more likely to be of the right disposition when the mare has been kept happy. Bringing the mare in at night can play an important part in this as the weather worsens. She will look forward to a nice dry, deep bed and her evening meal and will invariably greet your approach to her field with a welcoming whinny as she comes to meet you. I use a deep-litter straw bed for night boxes, and find that most mares use this regularly at night for periods of rest lying down. Although the odd night spent out in the open will do no great harm to a healthy, well-fed brood mare, a prolonged spell of cold weather with biting winds and lashing rain will make her miserable and pull her condition down. By bringing mares in at night you also gain the advantage of being able to look them over and examine them for injury at close quarters. It is a good idea to give them a quick rub over with your hand as they stand at the manger, as this keeps them well handled as well as revealing any lumps or bumps. I examine their udders, pull their ears, inspect their feet and generally keep them well disciplined, which makes a tremendous difference later on.

Make sure that night boxes are not draughty but leave the top door open to ensure a plentiful supply of fresh air. The box must not be allowed to become warm and stuffy, as if a mare is brought in to or turned out from a stuffy atmosphere she will be liable to catch cold or a chill. In frosty weather she will be best kept in until the early frost has gone because eating frosted grass can produce colic, and never turn her out immediately after a large feed of concentrates. In all but the most severe weather her need for some exercise will be best served by turning her out for a period during the day, and even snow will do no harm; though thawing ground that is frozen beneath will be slippery and dangerous.

4 Foaling

Mares carry their foal for eleven months – normally from 334 to 340 days, though as little as 325 days is not unusual and there have been cases where a mare has gone 400 days. If the mare is to remain at home for foaling, as the time draws near you will have to decide whether she is to foal outside or in a specially prepared foaling box. Either method can prove equally satisfactory, and your choice will depend on you own preference and particular circumstances. You should contact your vet to let him know the approximate date he may be required if things go wrong, and make enquiries about the nearest service that can help you should you be unfortunate enough to lose your mare and be left with an orphan foal. This event will present an emergency in which you will need to act quickly if you are to save the foal, and any advance information you have acquired will save valuable time. Should the worst happen your vet may be able to put you in touch with someone fairly near whose mare has recently lost her foal, and you may be able to save yours by introducing it to a foster mother. If not, there are services that can get a supply of mare's milk to you containing the essential colostrum which the foal must have for the functioning of its system in the early hours of its life. In the UK the National Foaling Bank, Newport, Shropshire, tel: Newport (0952) 811234 will help you, and there are several retailers nationally that can supply 3 kg containers of foaling milk that you can purchase beforehand as a standby. It is also advisable to have a bottle feeder on hand. Some people after giving an initial feed of foaling milk have managed to rear an orphan foal by bottle feeding with goat's milk, but the incidence of death in orphaned foals is high and raising them is a tricky business requiring dedication and luck.

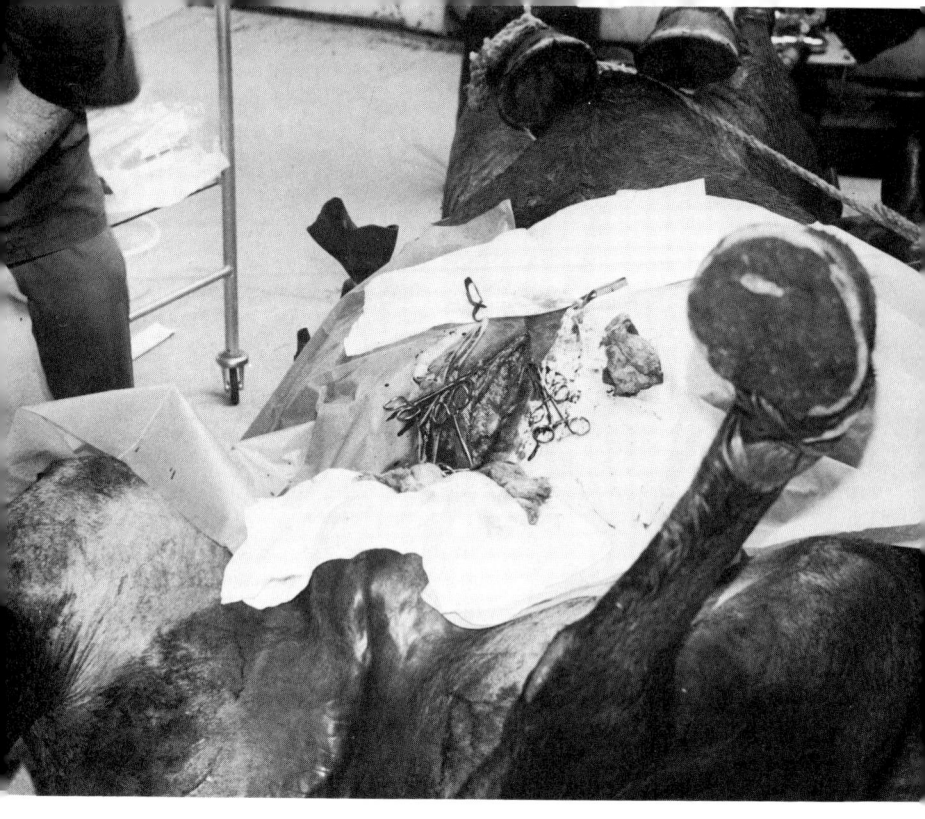

A fully anaesthetized mare undergoing a caesarean operation.

Fortunately, any problems at foaling time that cause the death of the mare are fairly rare with the benefit of modern veterinary assistance, and the vast majority of mares need no special help and have no difficulty. As to the well being of the foal at and immediately after birth, most mares will manage this well enough if left to their own devices provided you have seen to it that there are no physical encumbrances to prevent this. I have already explained the practical precautions that should be carried out to ensure as far as possible that no accidents happen – safe boxes of adequate size for those foaling inside; level paddocks without steep banks or ditches for those left to foal outside.

Most mares (but by no means all) will begin to show signs of the impending birth as the time approaches. The udder will start to enlarge; you may notice that it goes down after exercise and is then enlarged again the following day. As her time gets nearer the udder will probably remain in the enlarged condition. Another sign is the shiny wax-like substance that appears around the enlarged teats of the udder; this is known as waxing and is usually an indication that foaling time is near. However, it caused me some concern on one occasion by appearing in May on a mare not due to foal before July – who then continued normally until her full time. A more reliable sign of approaching birth is when the mare's milk drips from her teats and can be seen on the inside of her hind legs. This is more likely with older mares having their second or subsequent foals; in young mares having their first foal there are often no such signs and the milk seems to enter the udder immediately after foaling.

Impending parturition is usually accompanied by a slackening of the muscles on either side of the tail and a lengthening and swelling of the vulva. The mare may sweat and become excited and restless, pawing her belly and generally acting uneasy, sometimes getting up and down frequently. On the other hand she may make little or no fuss beforehand, appearing quite relaxed one minute and presenting her foal the next. If the mare shows any or all of the signs described but goes on for several days or even weeks after this without foaling it would be wise to expect trouble – and even twins! On the other hand, it is easy to be caught out by the mare foaling earlier than expected. I had one mare that I put into foal as a two-year-old and watched very carefully as her time approached. I examined her one night when I fed her in the paddock about a fortnight before she was due to foal and was convinced that she

This foal was too large to go through the pelvic arch and the mare had rotated the uterus by her continual heaving.

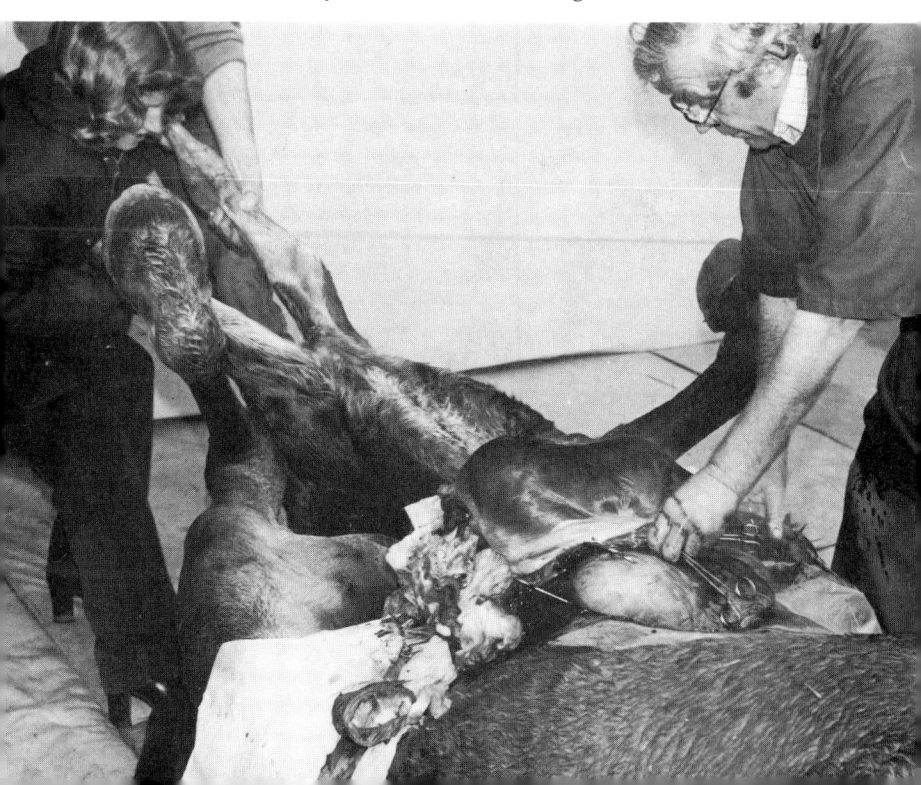

was nowhere near foaling – in fact I began to doubt that she was in foal at all. That night, during a thunderstorm in the early hours, she gave birth to a normal, healthy foal that I found standing beside her when I went to feed her first thing in the morning. She had only received a service from my stallion during one brief heat the preceding year and had only gone 325 days from the service. It shows that none of the signs referred to need be present prior to foaling and just when you think you know what to look for they can make a fool of you.

The actual birth

If you foal your mare inside and sit up to witness the event you will need to check her at half hourly intervals throughout the night as her time gets near. When she is about to foal you will probably see her walk round the box pausing restlessly here and there, and then quite suddenly things will begin to happen. Her tail will lift and brown water will be seen to flow from the vulva, followed very quickly by a white transparent bubble that is filled with clear fluid. As it protrudes you should see the front feet of the foal inside; this is the normal position with one foot slightly in advance of the other, ensuring that the foal's shoulders are angled correctly for their passage through the pelvic arch. If she has not already done so the mare will then get down and the foaling will be completed very quickly. The mare will normally remain down for about twenty minutes or so. The transparent skin of the amnion will often burst as the foal's feet poke through, or it may remain intact until the foal is completely free of the mare and then be broken by the struggling foal. It is vitally important that this protecting skin (allantois) is clear of the foal's head once it is free as otherwise it can suffocate.

Sometimes, after the foal's front feet and head are free from the mare and have broken through the bag, the remainder of the foal will remain inside the mare who will then lie at rest before straining to deliver the back end of the foal. This is quite normal, and the mare should be left for a time to complete the birth naturally. If, however, after half an hour or so the mare is obviously having great difficulty in completing the birth the foal can be assisted out. This is best done by an experienced handler as it can be dangerous to the foal, whose spine could suffer injury. Should you need to do it then grasp the foal's front feet and pull firmly and smoothly in rhythm with the mare's straining efforts to make the delivery. Take care not to break the umbilical cord at this stage as the foal will be receiving a considerable amount of blood through it at this time. When the mare has rested sufficiently she will get up, breaking the cord naturally. This will often leave a length of cord several inches long hanging from the foal, and

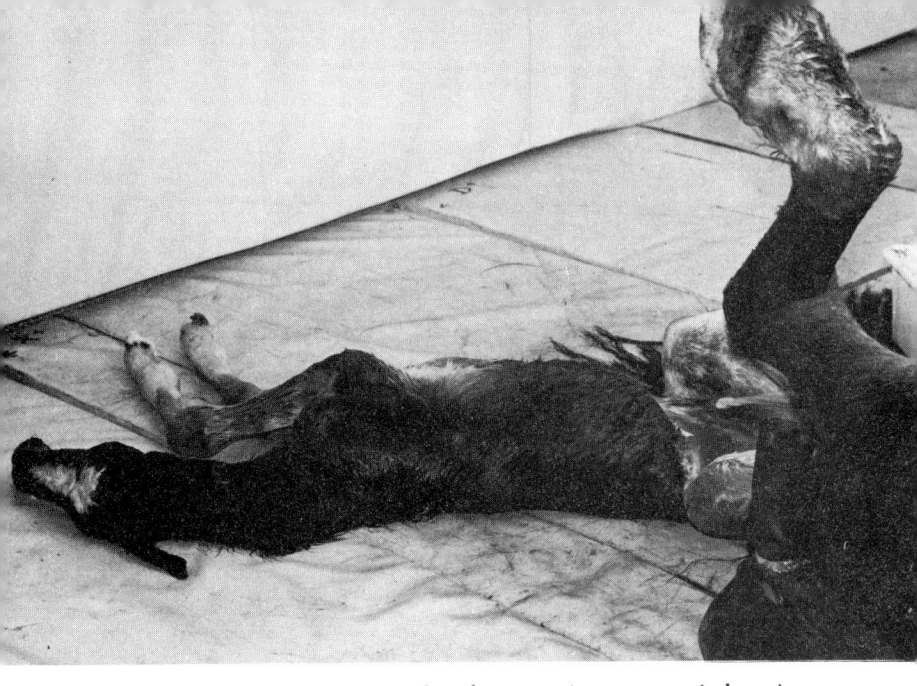

The foal had died inside the mare but the operation was carried out in time to save the mare's life. The unbroken umbilical cord can be clearly seen below the mare's hock.

A natural birth, in which the newborn foal is free from the enveloping membrane of the placenta but is still connected to the amnion with its hind legs still inside the mare. The foal continues to receive blood and oxygen through the unbroken umbilical cord at this stage.

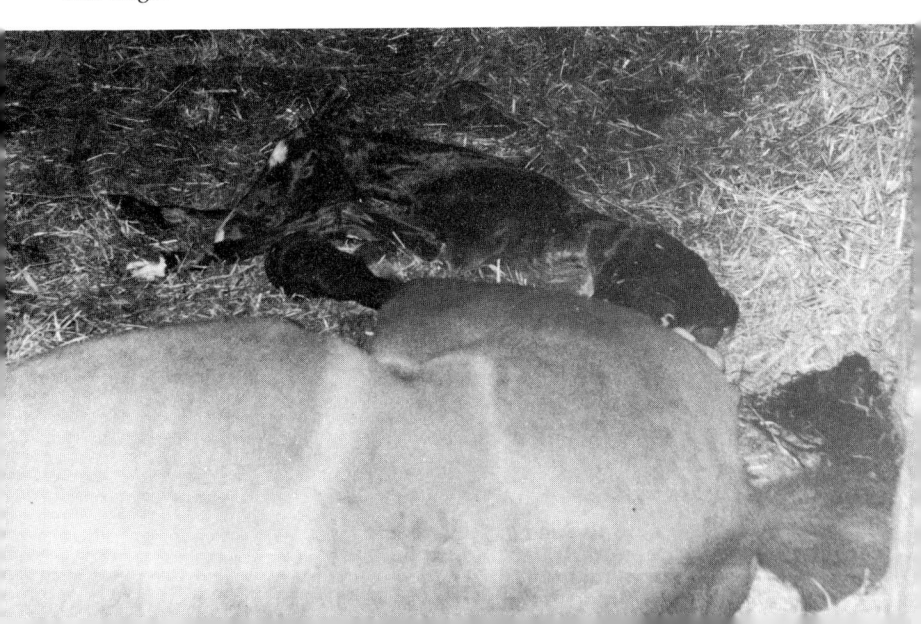

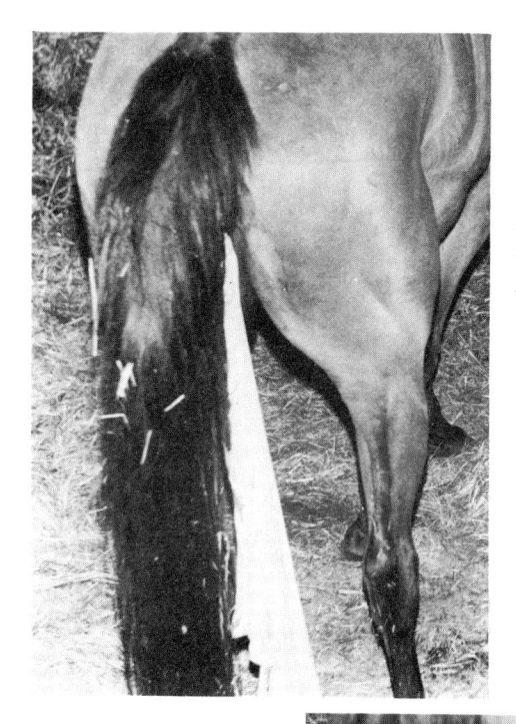

As the mare stands up the foal comes free, breaking the umbilical cord and leaving the afterbirth hanging from the mare's vulva.

The afterbirth (known as the placenta while it contained the foal inside the mare) now drops free from the mare. The transparent part that contained the clear fluid (allantois) and the darker part of blood vessels (amnion) can be distinctly seen.

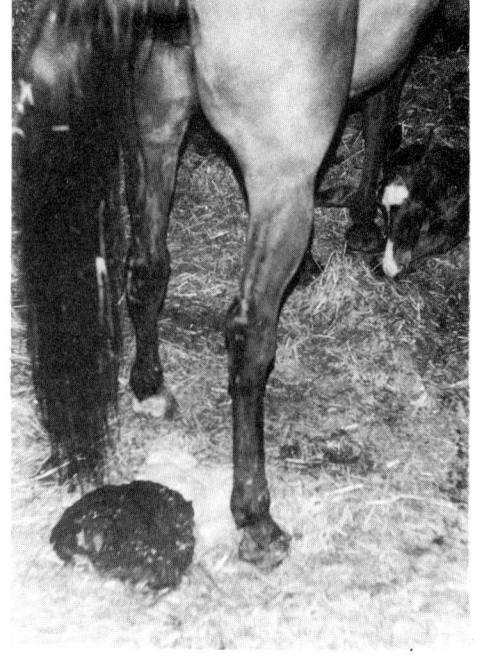

though this looks rather unsightly it will gradually disappear with time. It may take several weeks to shrink and be taken up in the skin of the growing foal but unless there is a hard lump in it indicating a possible hernia there is nothing to worry about.

The wound left by the ruptured umbilical cord will seal itself naturally, but as it is a source of infection to the foal at the time of breaking it should be sterilized without delay. Iodine powder used to be used for this, but it is now more often carried out by using a 'Violet' spray. This can be obtained from your vet and is widely used to prevent infection in superficial lesions.

After she has foaled the mare will lick the foal to clean it, and more importantly to stimulate its circulation. The temperature inside the womb is much higher than it is outside; the licking will activate the foal's circulation and help to keep it warm in the initial stages of its life in the outside world. Sometimes, and this is especially true with very young mares having their first foal, the mare will be frightened by the act and sight of foaling and will immediately spring up, rupturing the cord. This will cause blood to pump out of the amnion as it hangs from the mare's vulva, making it appear that the mare is haemorrhaging. In fact this is the blood that the foal would have received through the cord had it not been severed prematurely, and after a short time this will stop naturally.

The mare looks on as the foal's navel stump, formed by the breaking of the umbilical cord, is sterilized.

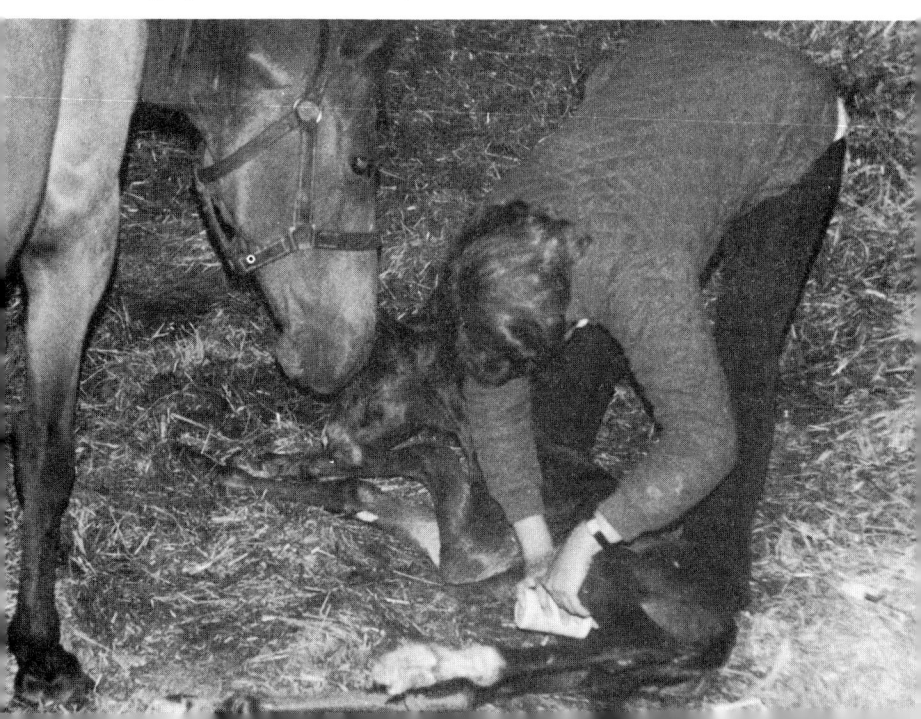

The loss of this blood will do no great harm to the foal in most cases as it will have received some during birth, but it is important to see that the foal suckles fairly soon as it will need the essential colostrum from the mare's first milk in order to sustain its life. The milk that the mare produces during the first day or two after foaling contains a high proportion of antibodies that are vital to the foal, who relies entirely on this fore-milk or colostrum for its immunity to disease and resistance to infection. The foal will absorb the colostrum through its gut into its bloodstream after suckling, and will thus be provided with protection against disease-bearing organisms. When it is about two days old the foal's own system will produce antibodies but it must receive initial protection from the mare; a day or two after parturition (birth) the mare's milk will differ in composition from that secreted at and immediately after the birth of the foal.

If the mare completes her foaling lying down and does not get up, thus breaking the umbilical cord, it can happen that the afterbirth comes completely away from her while she is still down. When this happens the cord will often be broken satisfactorily as the foal struggles to get up, but if the foal also remains resting it will become necessary to sever it. To do this it must be tied in two places with twine and then cut between the two ligatures. Leave about 1½ in (4 cm) between the foal's abdomen and where you make the first tie and the same distance between the ties, and immediately sterilize the navel stump after severance. It is crucial that you remember that once the cord is severed the foal will no longer be receiving the oxygen that was present in the blood flowing through to it, so its own breathing must be fully established. This is why it is absolutely vital that the foal's head is clear of the enveloping membrane of the amnion.

In order for the reader to know exactly what is being described I have included photographs of a discarded placenta, or afterbirth as it is commonly called. It comprises an outer transparent membrane (allantois) and a main body (amnion) which supplies the foal with nourishment through the umbilical cord that joins them. It is usual, after the foal has broken out, for the whole placenta to separate from the mare as one slimy mass of tissue, membrane and blood, but the colour photos I have included show the two quite separate parts exactly as I found them outside from the mare I described earlier that foaled after 325 days.

The whole afterbirth should separate from the uterus and clear completely from the mare within a few hours of foaling; if it remains hanging from her and has not been expelled after six or eight hours a vet should be consulted because of the danger to the mare of complications from septicaemia, which will have fatal consequences. Some afterbirths are ejected during the foaling process, but those that remain partially in the mare and hang down from her vulva should be tied up to her tail to await natural evacuation; this will eliminate the danger of her accidentally treading on it and possibly causing it to be parted forcibly, which would also cause complications. When it is finally ejected, spread it out and examine it to make sure it is all present and none has been left inside the mare. When the membranes are spread out you should see two distinct 'horns'; the pregnant horn and the non-pregnant horn forming a complete bag. The photographs will show what to look for, but if you are in doubt – and especially if the mare shows signs of distress or appears unwell – then a vet should immediately be sent for to examine the mare.

If, as is usual, all is well and mother and offspring are taking an interest in each other and are generally happy and contented, collect up the afterbirth and either burn or bury it. If the foaling has been carried out inside you should also remove and burn the soiled straw bedding and put down plenty of fresh, clean straw. Shavings can be used, but most people find clean straw the best and most satisfactory bedding medium at this time.

After a successful birth there are three things to make sure of before you can rest assured that all is well: the foal must be seen to suckle; it must pass some dung; the act of urinating should be accomplished.

Suckling

The foal should be on its feet and suckling within half an hour of birth, or at most by two hours after. If it has not suckled by this time it must certainly be helped and encouraged to do so. Though one need not be too worried at this stage, the mare's early milk is important not only in protecting the foal from infection but also to set its own body functions in motion. Most often one is surprised at the persistence of the newborn foal as it struggles to its feet and searches for its mother's udder. It will be very unsteady on its feet, staggering about around the mare, and after feeding will soon lie down again. After a few feeds in this way it will quickly gain strength and delight you with its perkiness; watching a foal in these early stages of life will imprint a picture on your memory that you will always remember with joy. It is sometimes necessary to guide the foal to the right position, for although the milk will often be running freely from the teats the foal may be unable to locate the exact place to feed and will try to suck at the mare's elbow or stifle. If no human guidance is at hand, however, the mare will invariably come to the foal's assistance by pushing it until it finds the right position.

Problem mares

There are occasions when although the foal is able and willing to suckle the mare will not allow it to do so. This may be because her udder is tender or, especially with very young mares, because she is nervous and not sure what it is all about. Sometimes the mare will positively reject the foal; when this happens you have a problem, as you must somehow get the mare to take the foal and this can be difficult and dangerous. I had a young mare that reacted in this way, not only refusing to allow her foal to suckle by kicking it away when it tried to do so but also lashing out violently at anyone trying to approach her. Normally a well mannered mare, she became very jealous after foaling, and being of Arab blood was like quicksilver when showing her displeasure. I believe it is true that about half the mares that reject their foals in this way are Arab. My mare had foaled outside in the paddock, and when we observed what was happening we tried to catch her up but her violence made this impossible. In the end we had to drive her and her offspring along the hedge and out through the gate after lining people up to drive her in the right direction so that eventually both she and the infant could be herded into a loose box. It took two hours, by which time the little foal was very weak and tired. With them both safely inside my assistant and I tried to coax the mare into letting us get near her but to no avail – approaching her head resulted in her swinging round and lashing out with her hind feet. It was soon apparent that we were risking not only our own lives but also that of the foal because of the mare's violence. Luckily she was wearing a headcollar and my helper, remembering an old gypsy trick she had been taught as a girl, went off for a walking stick. When she returned with it she used the crooked handle to hook the mare's headcollar from outside the box, and gently persuaded the animal to come around so that she could be held. We then went inside, and I held the mare and picked up a front leg to prevent her from kicking my assistant, who tried to introduce the foal to suckling. The mare would have none of it, and leaped and reared with me around the box. After several attempts it had to be abandoned. Success was finally achieved by my twitching the mare and holding her head in the corner with a front leg held firmly up beneath her chest, while my assistant held the foal's mouth to suckle from the mare. It took six hours in all, by which time the foal was exhausted and could not stand, being almost too weak to suckle. Every hour or so this exercise was repeated; by the end of the day the mare had returned to a more normal frame of mind and would let the foal suckle freely without restraint. Her jealousy continued for two or three days, after which she would allow anyone in the box. Fortunately this does not often happen, but if it does and you cannot get the mare to accept the foal then an alternative such as a foster mother or bottle

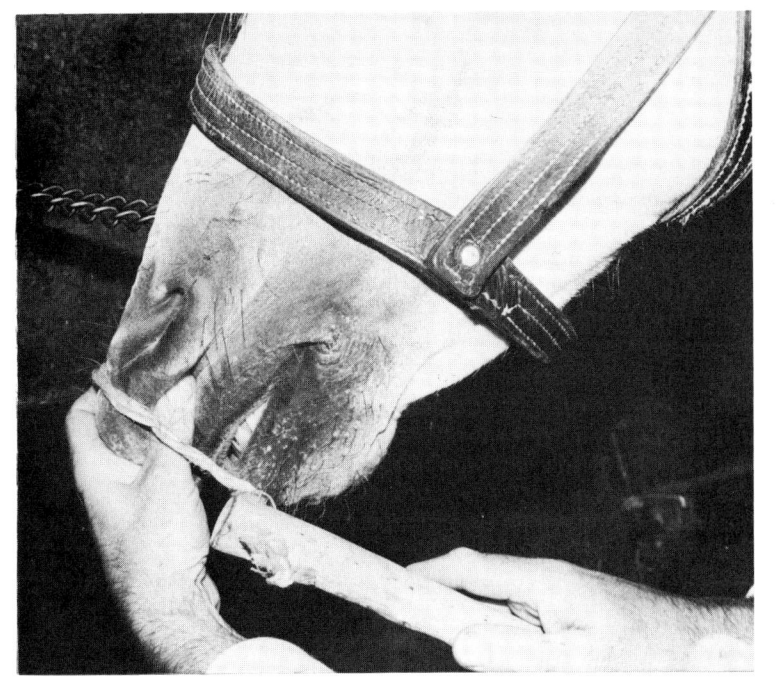

Twitching can occasionally be needed, as explained in the text (problem mares, Chapter 4): slip the cord over the fingers that are grasping the upper lip, and tighten the cord by twisting the handle.

Pinch enough of the upper lip to prevent the cord slipping off, but keep it below the nostrils to allow free breathing. Have the knot and handle at the side and keep it clear of the teeth and gums to prevent injury.

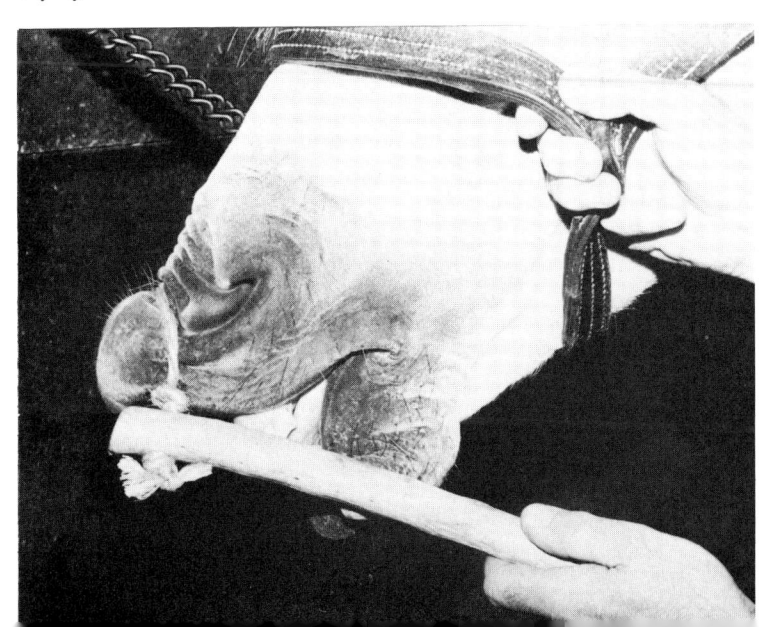

feeding will have to be arranged. We put an alternative plan in action, but there was no mare in the area that had recently lost a foal; my vet could not come for several hours to sedate the mare as he was engaged with an operation on a cow; we obtained a supply of goat's milk, and my assistant courageously said she would try to milk the mare for the vital colostrum. As it turned out we managed to avoid this, and happily for all concerned were able to persuade the mare of her responsibilities.

In the following days, to my delight and amusement, the little colt foal was such a fighter he would push himself across the floor with his hind legs and then get to his feet by pushing his head against the wall to struggle up to a standing position. He soon made up for lost time, and as his strength grew he would punch his head up at the mare's udder as he suckled, almost lifting her off the ground as though he was getting his own back! The mare tolerated this, and merely glanced back at him with approval. As the foal was sired by my Andalusian stallion we named him 'Luchador', Spanish for struggle – which is what it had been for all of us.

If you should be unlucky enough ever to have a similar problem then my experience related here should prove helpful. It is a job for two people – one person alone cannot manage, and more than two will get in each other's way and make things more dangerous.

The foal's first dung

The next important function to observe is that of the foal passing its first dung (meconium) within an hour or two of foaling. Failure to do this will need attention, as complications may arise that if left unrelieved can lead to the foal's death. Its first dung will quite often only be evacuated after some straining, and will consist of little faeces like hard, dark balls about the size of a sheep's dropping. This constipation is more likely in foals born of stud-kept mares that have been receiving concentrates; foals born to mares out at grass usually have less trouble. If the foal is seen to strain constantly without success and a vet cannot be obtained it may be possible for you to effect a remedy yourself if the trouble is merely constipation in the end of the rectum. Have an assistant hold the foal, then oil or Vaseline one finger and gently probe its rectum to try and find the cause of the obstruction. Care must be taken as over-zealous probing can be dangerous, and anything more than gentle easing near the entrance should always be done by a vet. A dose of castor oil or liquid paraffin can be given once the foal has suckled and this may help the condition, but continuing discomfort will need to be dealt with as described in chapter 9.

The third thing essential to the foal's well being is that of urinating. This will normally occur without straining and rarely presents any problems. If you are not present at the time of foaling and only see the mare and foal some hours after, look around for signs of the foal's dung and stay with them long enough to witness all the functions described. Only if either are showing signs of distress need you worry about abnormalities; if both are happy and at ease you can safely assume all is well as you settle down to watch them.

After foaling – the first days

If the mare has been foaled inside there is no reason why she and her foal should not go out onto grass the next day provided that the weather is suitable. Once the foal has suckled normally and is strong enough to move about, the sooner they both get used to being outside the better. Keeping them both cooped up will often lead to unwanted mad cavorting about when freedom is eventually given. Obviously the weather needs to be fair, as the foal will spend a great deal of its early life lying down and there is little point in having it exposed to wet and windy conditions at this time. For those born out in the open there is no reason why they should not remain out day and night if the weather is warm and dry. This is another reason why I prefer foals born in July and August, as the nights are much warmer for them to be outside whereas spring nights can be very cold even after warm days. However, late foaling will not be so advantageous to those needing to breed from the mare again straight away, as there will be less time left for her to come into season, and to cover her again much later would produce a very late foal the following year which would be less convenient. Not all mares oblige by coming into season again after foaling, and some even miss for the remainder of that year, but mostly they show signs of oestrus about a week to ten days after foaling, and this is known as the foaling heat. Some people use this period to have the stallion serve the mare again, but I prefer to wait until the following oestrus; the foaling heat is low in fertility and has a success rate of only some 25 per cent; by the next heat the mare will have cleaned properly and her fertility will be much higher. During the first week after foaling you may notice small amounts of brown-coloured blood dripping from the mare's vulva or see stains of this colour high up on the inside of her buttocks. This is the normal discharge as her organs clear the waste materials left from the act of foaling, and there is no need for concern unless it is very pronounced and continues for longer than the first week.

The foal – first lessons

Within a day or two of birth, and certainly by a week old, I like to have the foal in a headcollar, or foal slip as it is known. These can

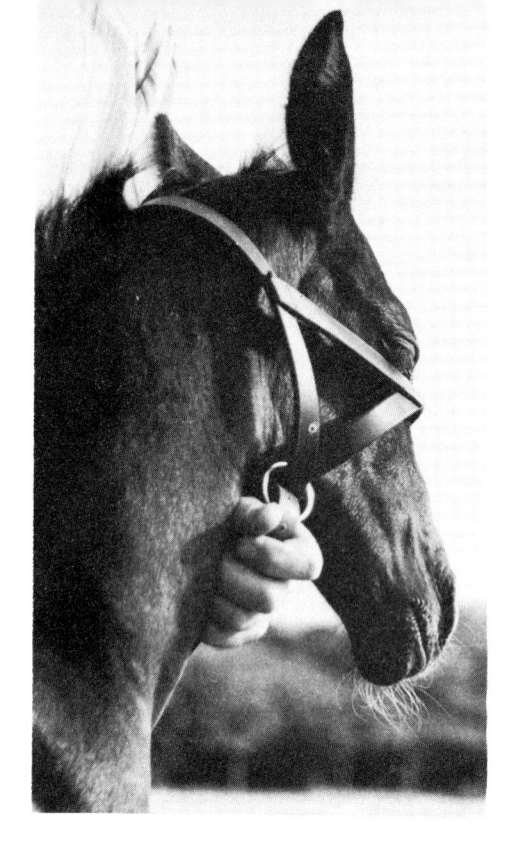

Within a few days of birth a foal slip (head-collar) should be fitted; make sure it is adjusted so that it does not interfere with the foal's eyes – the photograph shows it buckled too high.

Properly adjusted, the foal slip presents no problems even when the foal suckles.

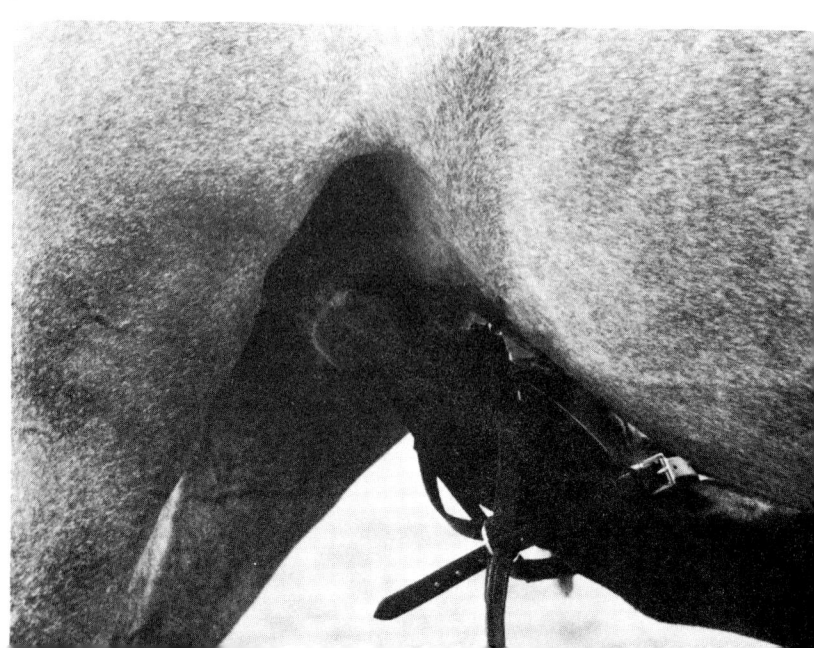

be made from nylon, but for comfort and for buckles that give better and more positive adjustment I prefer a good leather one. Make sure that it is comfortable, and adjusted to be large enough not to restrict the little animal by rubbing or interfering with its eyes but not so loose as to be a nuisance by flapping or dangling about. When you fit it on the foal's head be calm and gentle and reassure the foal. It is easier to do this in a loose box than outside. Only leave it on for a few hours each day so that it does not rub sore patches on tender young skin, and also so that the foal gets used to it being put on and off. If you fail to get the young foal used to this early on it will present problems later when it will be head-shy and much stronger, causing trouble in catching and holding it. Always be patient but persistent with difficult foals, and if a battle has to be fought make sure that the foal does not win; this is essential if you are not to raise a headstrong, unmanageable animal. Don't be too discouraged if for a while little progress is made towards trouble-free handling as it will come in the end, and in many cases a nervous, shy or headstrong foal will quite suddenly learn to give up the struggle and then become the friendliest of all. From the first week onwards it is also good practice to handle it all over and gently pick up its feet, teaching it to stand quietly balanced on three legs. This obviously needs to be for brief periods as weak joints need to be handled carefully and strain avoided. These first lessons in handling and discipline are most important and will save much time and trouble later. Do not make the mistake of treating the foal like a pet by fondling it foolishly and so spoiling it. Remember the object is to teach it manners and respect, and in return give reward and kindness. Sentimentality and affection such as that existing between owner and dog, for example, has no place in the rearing of horses, who are incapable of understanding this kind of relationship.

5 Raising young stock

The rearing of young stock to become well-mannered, healthy horses will depend on three things: the health and vigour of the foal at birth; feeding, both by the dam's supply of milk and your supplementary feeding before and after weaning; careful and progressive handling and training right up to and after breaking.

In dealing with the first of these your pre-natal feeding and management of the mare will play a decisive part, as will your skilful selection of the parents, but beyond that it will be outside your control and one must rely on nature and good luck. I shall deal with the second in chapter 7, which includes the feeding of brood mares as well as young stock. In the third category I have already explained why I consider early handling and discipline to be necesssary from the outset, and I will now enlarge upon how this is built upon as the foal grows in age and strength.

During its first few days of life a foal will show incredible progress in strength and agility, and after only a week or two its leaping and cavorting about in inquisitive and mischievous antics will give much pleasure and amusement. It can sometimes be quite worrying to watch it showing off for fear that injury might result, but as long as the paddock is not too sloping and is free of ditches or sharp obstructions a foal is unlikely to do itself harm. It must by nature be allowed to develop and grow in this way, and these sudden bursts of energy will be followed by long periods of rest lying down in close proximity to its dam. I shall never forget the joy of witnessing a particular performance by a two-week-old filly foal that I bred. We stood bemused and afterwards all agreed that the display was almost like a rehearsed performance of advanced dressage and high school, including the airs above ground. These were carried out in one direction and then after a slight pause and a glance at us to see we were still watching she did them in the other direction! She is still but a yearling, the product of my Andalusian stallion and a mare of unknown quantity, so I can only hope that her antics hold out a

When teaching a foal to lead never try to pull it along by standing in front of it, as this will only result in greater resistance and straining backwards.

promise of what she will be naturally gifted with in the future.

Having successfully introduced the foal to a headcollar it is time to clip on a lead rein and teach it to walk beside you. At first it can be helpful to have someone lead the mare in front as the foal will naturally follow along behind. When the mare is moved about by being led from place to place always see to it that the foal is led along as well, and this will help it learn to lead without trouble. After this try leading the foal away on its own, making a circle around the mare so they can both see each other. As time goes on gradually enlarge the circle so that both of them get used to being separated by a greater distance. Resistance will bound to be offered with the first attempts, but never get in front of the foal and try to pull it along as this will only result in greater resistance and straining backwards. More dangerous is that the foal will then invariably rear up and half turn away, which will cause it to be in great danger of being pulled over backwards. If this should happen to you, always respond immediately by stepping towards the foal and slackening the rope, as to topple it over onto its back is to be avoided for obvious reasons. The way to get it to lead properly is to hold the rope at its chest and place your free hand behind its quarters, encouraging it forwards without pressure on its head. It will soon learn to relax and walk forwards, whereupon you walk by its side keeping a hand on its rump to keep it straight. Some foals are slower to learn than others, but calm, persistant encouragement from you will teach every foal this compliance in the end. It is as well to make the point here, though, that even young foals can be surprisingly strong and the teaching of youngsters is no job for the aged or infirm as considerable physical strength and agility on the part of the handler will be required.

Introducing lungeing

By the time the foal is two or three months old I like to begin what I call introductory lungeing as a natural progression from being led, which should now cause no bother. Using the foal slip with a lead rein of some 5–6 ft in length (1.75 m) I hold a riding stick in my free hand and as the foal walks on I stand still and encourage it to carry on around me. Once this has been learned and can be obtained without difficulty I begin teaching it to stop on command by a slight shake of the rope as I call the word 'halt'. This is further reinforced when necessary by changing the stick over to my other hand and holding it up in front of the foal. This training can be carried out in front of the loose box so that the mare can watch over the door and the foal will be aware of her presence. The foal will be more inclined to stop as it passes the door so you can ask for the halt then to teach initial compliance. When this is learned speak quietly to it, telling it to stand still as you walk up and pat it. Soon it will stand on its own, alert and ready to walk on again as you give the command and encourage it forwards, with a slight tap with the stick if required. After several weeks of this I begin teaching it to turn round after halting and walk off again in the opposite direction.

These training sessions only last for a few minutes, as on no account must the foal become tired or grow bored with it. If this happens it might damage its joints as well as building up a mental resistance that will sour it for future work. Naturally things do not always go smoothly and sometimes the foal's exuberance takes over, but when it does always remain calm and try to finish up with obedience to your final request so that it can be congratulated with a pat and returned to its mother. I find that if these little training sessions are carried out at feed time the feed can then be given to both mare and foal as a form of reward. I leave the other foals and mares together as I select each foal for its turn and they usually stand watching, which I believe helps them to learn the procedure.

Another practice I employ is to stand by the foal as it feeds at the manger, and when it gets accustomed to this I place my hand on its back for a while. When this is tolerated I slide my arm around its rib cage, and eventually reach the stage where I can

To get the foal to lead properly begin by holding the lead rope at its chest and place your free hand over its quarters to encourage it forwards.

At first it is a good idea to have someone lead the mare, allowing the foal to follow along behind while you walk by its side with a hand over its rump to keep it straight.

clasp my arms around it, linking my fingers together underneath. Finally I do this and also let the weight of my body and the restriction of my arms envelop the foal with some pressure. This is the first lesson in getting it used to the time when a girth will encircle it and be tightened. Later on, when it has finished its short period of introductory lungeing and is taken to the manger, I use the lunge rope to slip around its middle and let it feel the effect of this being tightened. These little lessons from almost the beginning of its life will be of immense value in the years to come when it must be taught to accept a girth and saddle.

First stage bitting

Another thing that can be taught as the foal develops is accepting a bit. When it is four or five months old I introduce it to a bit; this is done with a thick, flexible rubber snaffle that I hold in its mouth for a few seconds. Repeating the exercise once or twice a week I wait until the bit is accepted without fuss, and then I fix it to a webbing strap that I pass over its head and behind the ears and adjust on the opposite ring of the snaffle. Standing by the foal's side I see to it that the bit is maintained in the correct position for a minute or two, ensuring that the foal's tongue is under the bit as it chews with its jaw mouthing the rubber. This is most important as on no account must the foal learn to get its tongue over the bit, as the habit will then be repeated later with disastrous effects in training. After a few weeks, depending on how the foal progresses, I fix this rubber bit to the foal slip and adjust it so that it is held correctly high up in the mouth and then continue the rope lungeing already described. The rope, of course, is attached to the headcollar under the foal's chin and not to the rubber snaffle.

Grooming

Once the foals stand and feed at the manger with their dams I also take this opportunity to stand by them and run my hand over them. This gets them used to the feel of the body contact and also provides an opportunity to check their limbs for lumps, heat, etc. and generally make sure that all is well. I also use a plastic curry comb (one that is worn so that the teeth have rounded edges) to give the various parts of the body a brushing. This is not done enough to be called grooming but merely gets them used to the feel of it. Particular parts like the mane and tail can be brushed more vigorously as this will stimulate the blood supply to the roots of the hair and encourage strong growth, which personally I like to see.

The thing always to bear in mind is that these activities must be carried out in such a way as to cause no distress to the foal, nor should it be continually harrassed. You must constantly remind yourself of the need for patience, and keep the periods of instruction short. There is a danger that foals that are willing to respond and quick to learn may be taken by an over-zealous trainer beyond what is wise for them to do at this early stage of their development. Continuous daily sessions are neither necessary nor desirable, and a couple of short periods a week will suffice by varying what is done each time. It must also be said that none of this early training is strictly necessary, and many people buy youngsters that have never had such training nor even been handled or disciplined for several years. Such animals can and often are trained successfully by a skilled handler and go on to become disciplined and satisfactory mounts. However, I believe it to be undisputable that the correct handling and training of foals and young stock as I have described will be most beneficial and will be rewarded later by well-mannered animals that are a joy to work with. This is where small-scale breeding has its advantages, as it allows extra time to be devoted to this early training.

Care of the feet

Handling the young foal's feet and teaching it to stand while you pick up each foot will prove of great value as it becomes necessary to attend to its feet as it gets older. By the time the foal is three months old you should be using a hoof pick carefully to clean out the dirt from the little feet that are now developing, and by five months a rasp can be used to tidy up rough edges of horn and reduce the length of the toe where this is excessive. Only the underside should be touched, and care must be taken to ensure that only very little horn is removed and that the shape and overall angle of the foot is not affected. Have the blacksmith pay a visit at about six months to attend to its feet, and when your youngster is first shod later on in its life there should be no trouble. The farrier will appreciate this, as trying to shoe a three- or four-year-old that has never had its feet handled can be anything but pleasant.

Anti-tetanus injections

When the foal is six to eight weeks old it should be given its first anti-tetanus injection, followed several weeks later by a booster; the vet will arrange to do this and inform you when the booster is necessary. Up to this time the foal will have protection from the antibodies it received from its dam, provided of course that she was properly immunized. It is most necessary that you have this done as foals are particularly liable to tetanus and are very vulnerable to it. Some people also have the anti-flu injection given at the same time by having the vet use the combined innoculation called Prevac T. I only have the foal injected with the anti-tetanus vaccine, and wait until it

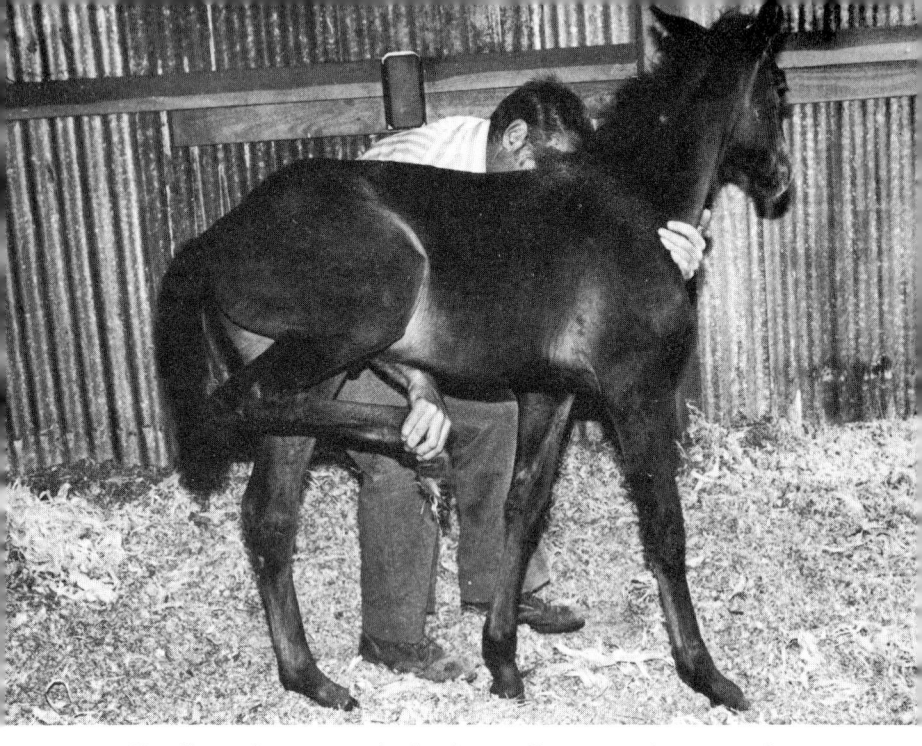

Handling the young foal's feet will prove of great value as it becomes necessary to attend to its feet when it gets older.

is between one and two years old before using the combined injection. This is only a personal choice; I feel the foal can be left to develop naturally its own resistance to bacteria other than tetanus, and I believe is helped by the fact that my mares all receive their regular flu jabs. It is worth mentioning here that some authorities are now prescribing that horses should have booster flu injections at more regular intervals than the twelve months originally prescribed. Some go as far as to say these are necessary every three months, but I hold to the view that once a year is sufficient. Owners who are persuaded to have their horses done more frequently than once a year should only have the flu vaccine repeated, and not use the combined Prevac T immunization, as extra tetanus protection will be unnecessary and could in some cases be harmful.

Worms

One of the greatest dangers to growing foals and youngsters, and one that is often not fully appreciated and may therefore be neglected, is that of worms. Foals and young stock are very susceptible to worm infestation, especially during the post-weaning period and their first autumn and winter; control of the parasites at this time is particularly important. Clean pastures and good feeding

Above: A mare will lick a newborn foal to stimulate its circulation. Make sure that the foal's navel stump, which is left by the breaking of the umbilical cord, is sterilized without delay. It will seal itself naturally but is a source of infection, especially to foals born indoors.

Below: Quite often the afterbirth does not clear the mare during the actual foaling, but it should come away naturally within a few hours of birth. To prevent the mare from forcibly removing it, with possible complications by treading on it, it should be tied up to her tail to await natural evacuation (Chapter 4).

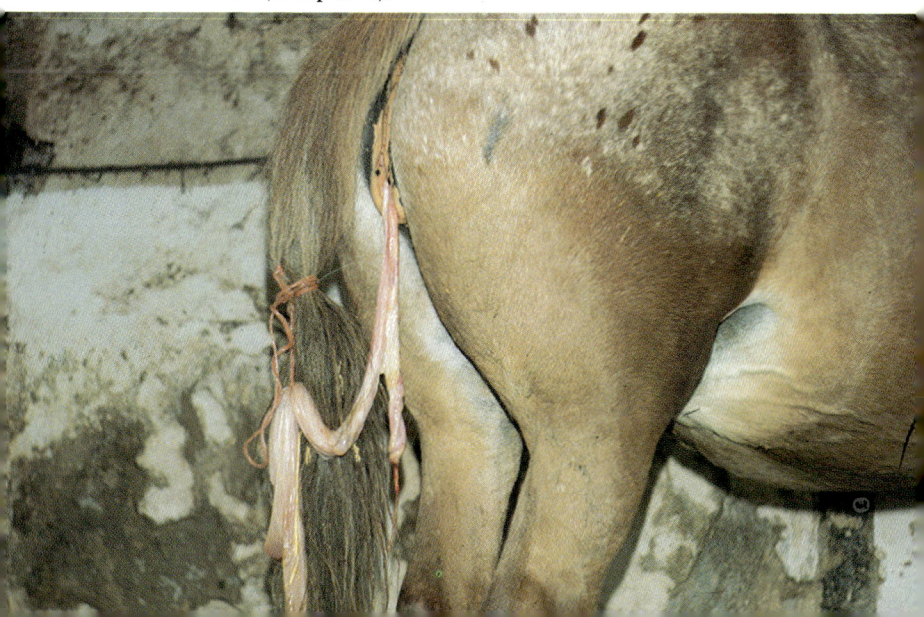

This afterbirth from a foaling out of doors had separated. Above: The outer transparent membrane (amnion) that envelops the foal and contains clear fluid. The grey object (centre left) is a calcium and salts deposit (hippomane) that is not often found. Below: The inner section (amnion) clearly showing the mass of blood vessels and broken umbilical cord (bottom left). The lower part also shows it to be complete, with two distinct 'horns' (Chapter 4).

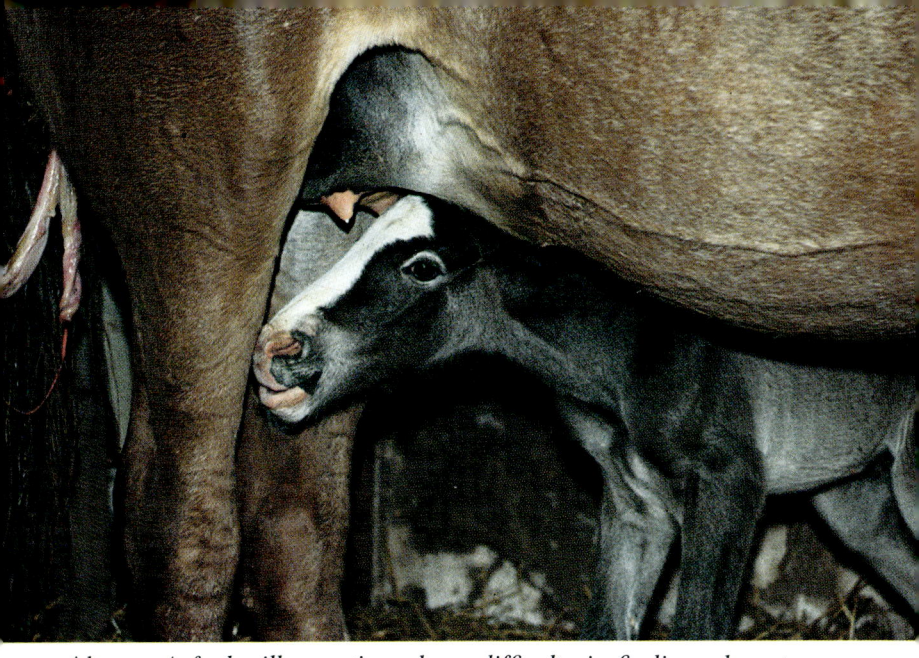

Above: *A foal will sometimes have difficulty in finding where to suckle. This mare's milk can be seen dripping freely from the teat but the foal has not yet discovered the source (Chapter 4).*

Below: *It may be necessary to help the foal to find the mare's udder, but usually the mare will direct the foal into the correct position.*

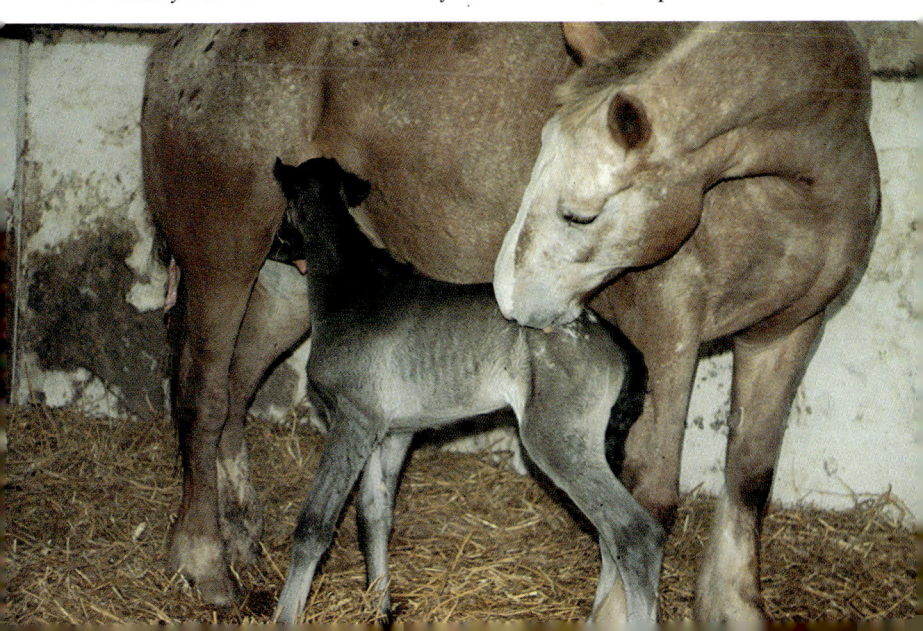

Above: *A mare that is fully in season can be safely teased by the stallion: here the stallion has worked his way along to nibble her flanks, causing the mare to straddle her hind legs and urinate (Chapter 8). Note the padding over the stable door to prevent the stallion rubbing himself.*

Below: *The oestrogen emitted in the mare's urine when she is in heat is detected by the stallion's olfactory nerves, causing his lips to expose his teeth, the top lip curling upwards to compress his nostrils; if this 'flehmen' is pronounced it is another sign of the mare's readiness (Chapter 8).*

Above: The mare should be allowed enough room to step forwards to adjust her balance when the stallion mounts; when properly inserted the stallion will usually 'flag' his tail and ejaculation takes place (Chapter 8). In the author's view two handlers are best when a mare is being serviced.

Below: His work done, the stallion is led back to his stable for hygienic cleansing while the mare is walked around for ten minutes to prevent her from straining and ejecting the spermatozoa. Note the tail bandage to prevent tail hairs being caught up during service.

Above: *Within a few hours of birth a foal will be up and skipping about, though still very unsteady.*

Below: *It will also be inquisitive enough to try its mother's grain feed, and will soon learn to eat a little.*

Above: *Foals spend a lot of time lying down (above) and will often remain motionless as you approach, giving the appearance of being dead.*

Below: *After only a week or two a foal will graze considerable amounts (below), and it is wise to have a well-fitting foal slip (headcollar) fitted by this time.*

Mother and daughter doing well – a sight to gladden the heart of any breeder. Successful breeding is the result of three things: careful selection, skilful management, and luck. This book should help anyone achieve the first two – the third is in the lap of the gods.

will help, as it is always the poorest fed that suffer most. Salt licks help the animal to keep down parasitic infestation by helping to prevent conditions in the digestive tracts that encourage the multiplication of ingested parasites. Mineral balance should also be maintained by including a mineral lick to ensure that there is no deficiency in the vital trace elements. Up to three weeks old foals can often be observed eating the mare's dung, and I believe this is in search of certain trace elements that it instinctively knows it needs. It is common sense, therefore, to ensure that the mare has been kept worm free (a relative term as some worms will be present in all animals) so that the worm larvae are not present in her droppings.

A very serious consequence of infestation of young stock from the large redworm (*Strongylus vulgaris*) and others such as *S.edentatus* and *S.equinus* is the resulting weakening of the arterial and intestinal walls. This is caused by migration of the worms into the bloodstream and penetration of intestinal mucous membrane and connective tissue. The precise route of migration and method of penetration is not clearly known and need not concern the average owner; what has been definitely established by autopsy is that extensive damage can be caused when the animal is young, and this weakness leads to a breakdown of the animal's vital organs such as

By the time the foal is three months old it should stand quietly without restraint as you handle it and pick up each foot.

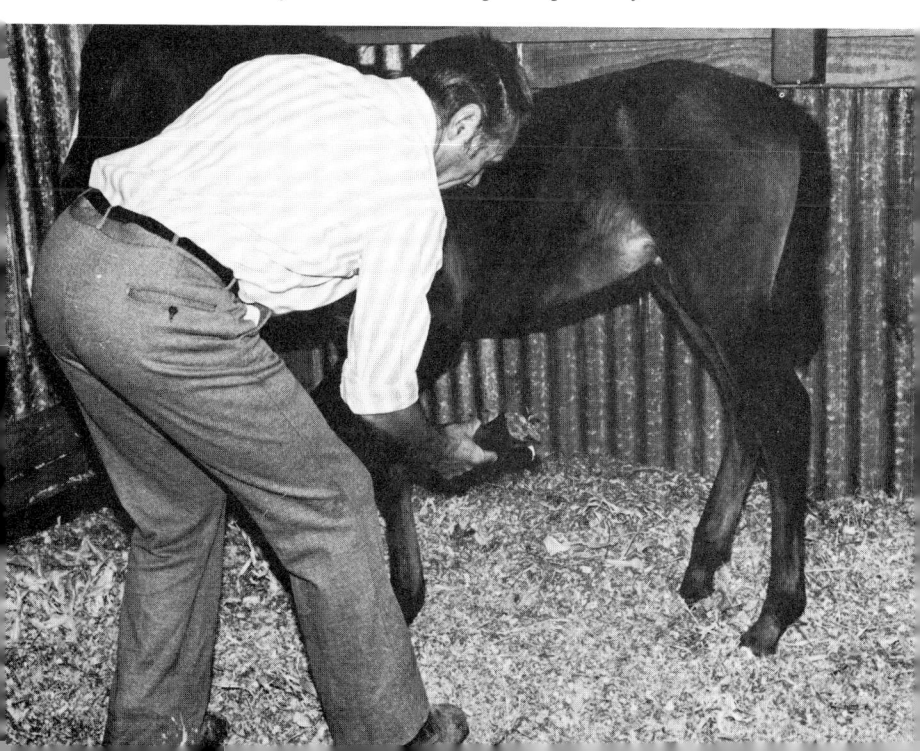

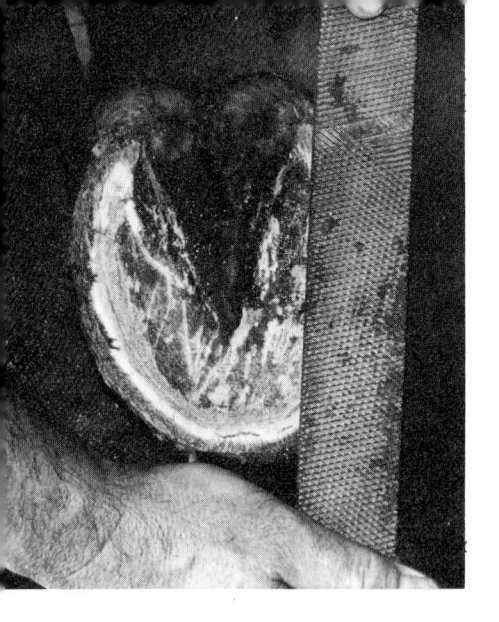

Above left: *As the young horse grows and develops the regular trimming of its feet will keep them strong and prevent splitting of the horn.*

Above right: *Toes should not be allowed to grow too long, and keeping the feet in good order as shown here will make shoeing much easier when the time comes.*

lungs and arteries when it is subjected to hard work later in life. One example is the breaking of blood vessels in the lungs of racehorses, which is now often equated with early worm infestation. Other sometimes fatal consequences such as puncturing of the intestines can also be traced to worm infestation at an early age.

Foals and young stock also have other worms to contend with, such as pinworms, tapeworms, ascarids, *S. westeri* and others, all of which should be dealt with by worming at more regular intervals than older horses. Different brands of powders and pastes should be used to overcome the likelihood of the worms building up a resistance to the treatments. I find it easiest to add powders to the feed of older horses, but I use a paste for foals and youngsters generally so that I know they get the full dose and it is not eaten by a more greedy feeder that shares the manger. These worming pastes can be obtained from your vet in handy plastic syringes that have the dosage marked off according to body weight. The first dose should be given when the foal is six weeks old and repeated at two- to three-month intervals depending on the number of stock grazing a given area. With good management the interval between dosage can gradually be lengthened so that worming is carried out at six-monthly intervals after the age of three years.

Winter care

It is difficult to give precise advice about how foals and young stock should be wintered, as everyone's circumstances will be somewhat different. There are several schemes that can be adopted; each depends for its success on your being able to recognize the needs in your particular case. The following categories will have to be considered: severity of the weather in your area; type of horse or pony; quantity and quality of feeding; location and amount of shelter.

In the UK the worst conditions will be early January through to March, and this period will increase in length and severity as one goes further north. Late-born foals will also be less well able to cope when winter comes, and it should be remembered that their first winter is all-important as their constitution will still be in the developing stage. The type that is being bred will need considering, as a hardy moorland pony will withstand severe weather more readily than other, finer breeds. Thicker skinned and shaggy-coated types will be much less susceptible to the cold and wet than horses of the 'blood' type. The feeding programme will play its part, and here it should be remembered that throughout its first year the foal will be making more growth than at any other time after that. It is debatable whether time lost in early growth is made up for later; again, I think this depends to a large extent on the breed. Tests were carried out on some Russian horses which were held back in their development by underfeeding when young, and they reached their full potential two years later than normal. The Spanish Andalusian has also traditionally been developed rather slowly on somewhat meagre fare, and is not considered mature until a year or two later than other breeds. However, no horse will grow and develop properly if it is substantially deprived as a foal, and with quick-growing breeds such as those with a fair proportion of Thoroughbred blood it is important to see they are adequately nourished during their first winter. Even with good feeding they will still fall off in condition if their need for shelter in bad weather is not met, so the location of their winter quarters needs to be considered. Unhedged fields on high ground will be exposed to the full blast of the bad weather, whereas those lower down and with thick hedgerows will be considerably more hospitable. A field shelter will make a tremendous difference, provided that it is not constructed so that it is draughty and waterlogged, which will discourage its use. The trouble, even with a good shelter, is that if one relies on foals or young stock (or for that matter any horse) to use it at will they quite often do not. Once caught out in wind and pouring rain they will stand huddled together in misery without the sense to go inside where a comfortable bed awaits them. You will soon learn that the winter months, and especially the first winter, are the most demanding and hard working for the DIY breeder. Unless your foals are of a very hardy type they will need to be brought in at night with their mares; after weaning the foals should be brought in at night and can be housed together; if you have only the one foal then it will have to be stabled alone. A field shelter can be used for this, but better is a large loose box or barn which will give adequate ventilation and a dry bed. Depending on the weather, youngsters can be left out at the beginning of the following spring, and their second and subsequent winters they can winter out provided that they are sheltered from the worst of the elements. Some will do better than others, and here you will prove your worth as a breeder: if you get your young stock through each winter still looking happy, round and solid and not bony, potbellied, weedy little miseries you can congratulate yourself on having mastered the art.

6 Weaning and afterwards

Weaning time will come when the foal is about six months old, and although it can be carried out earlier there is normally no point in this as the foal will continue to benefit from its dam's milk even when this is being supplemented by small grain feeds. There is a body of opinion that suggests weaning should not be carried out at all as it is more natural to leave the foal with its mother. It is argued that moorland ponies and semi-wild breeds are left together and the mare's milk will dry up naturally at about ten months after foaling. When this occurs the mare will prevent the foal from suckling if it still persists in trying to obtain milk from her. The advantages in this are said to be that not only does the foal benefit from the milk for a longer period but also it will not suffer the trauma of sudden separation from its mother at a very early age.

Separation will cause psychological problems, and although I do not agree with the view that the foal's whole future behavioural pattern is adversely affected by separation, I do agree that some disturbances can and often do take place. After separation the foal can develop a bad temper, start to crib bite, become less tractable and be more difficult to handle. However, I think this is more likely with foals that have not been well handled prior to weaning, which is another reason why time spent with them from the beginning pays dividends. Single foals will suffer from this isolation much more than two or more that can be weaned off their dams and kept together.

Another advantage of keeping foals and their dams together for a longer period is that they learn from the older mares and are disciplined by them, whereas if they are weaned early and turned out together they can become very boisterous and bossy with their human attendant. Brought up with older mares and in family groups the natural discipline of the herd will be learned and they will be much less likely to 'try it on' and take liberties. For the small-scale breeder of perhaps only a single foal I think it best to adopt the six-month period for weaning, keeping in mind the inevitable problems and trying to make the foal's transition from dependence on the mare to individual status as smooth and untraumatic as possible.

When the foal is weaned it will have to be taken right away from the mare so they are out of seeing and hearing distance of each other. It is best to use strong loose boxes for this, or somewhere that they can be shut up without risk of their injuring themselves by trying to jump out. Generally this shutting away will last from a few days to a week, after which time they can go out as normal though not of course together as they should be kept a safe distance apart for quite a while yet. Initially it is wise to keep top doors of loose boxes closed and leave the foal with hay and water. They will by this time also be eating grain, so that small feeds can be given as described in the next chapter. Always be sure to double check that the loose box used for the foal at weaning has solid sides and is free from obstructions that could cause injury. Quite often a foal will at first go crazy, and I have known them to climb up the walls or try to smash through a window; when this happens it is best to 'black out' the box by covering over any windows or openings until the foal calms down; there will also be a tendency for them to sweat up for a few hours after separation.

The mare will need to have her rations cut down somewhat for a few days, needing to have only hay and her water restricted to about half what she has been drinking so that her milk dries up. Mares that make a lot of milk may get trouble at this time with a hard and distended udder, and in this case it may be necessary to draw off a little by hand. When this happens only take away enough milk to ease the mare if she seems to be in pain, as to continue milking will only encourage more milk production. Normally if weaning is left to six months there is no trouble and after about a week she will be ready to return to full rations and go out for grazing.

Foals – colts and fillies

When breeding more than one foal you may be lucky and produce, say, two of the same sex, who can be kept together after weaning without trouble. More probably you will get one filly and one colt foal, which will create problems. Two or more of the same sex will live and do well together, but one of each sex means either keeping them apart, making twice the work, or having the colt 'cut' rather early. By the age of four months a colt foal will be seen to bite at a filly's withers and even try to mount her. This is done in a playful manner, and romping together in this way will do no harm – in fact it will help the filly to understand later what is happening if she is bred from. However, time passes quickly and by one year old a colt will be man enough to do the job properly, and a filly of that age will be sufficiently well developed to be got into foal. This of course is most undesirable, and will need to be prevented. It is difficult to be precise about individual cases of sexual development, as a colt's testicles can be in the scrotum at birth or may not arrive there until much later. This can make castration difficult if it is undertaken too early; many vets prefer to wait until a colt is a year old as they then have more to work on! Some people also consider that to have a colt castrated too early reduces its chance of reaching full growth, while others feel that to delay it alters its character and makes it bossy later in life. Neither of these views is strictly true in my opinion, as character and growth are determined by other criteria such as management, feeding and inherited tendencies. Many successful geldings have resulted from foals that have been cut before they were six months old, and many others that were castrated a month or two before being weaned. On the other hand it is equally true that many are left longer with results that are just as satisfactory. If you intend to keep a filly and a colt foal together your choice will be determined by necessity, and you will need to have the operation carried out sooner rather than later. Your vet will have to be consulted and will advise what is best in your particular case. You can then decide whether to take the colt to his surgery to have it done under general anaesthetic, or whether he should come to you and use a 'knock-out' drug to perform the castration. If the weather is fine and mild the latter can be done out in the open in a level paddock where there will be less risk from infection, and healing will also be better and quicker if the foal can exercise itself immediately afterwards. This is dealt with fully in chapter 9.

Teeth

An adult mare has thirty-six permanent teeth, the upper jaw having six incisors in the front and six molar teeth along each side; the lower jaw has the same. The adult male horse has forty permanent teeth; these consist of thirty-six as described plus four canine teeth or 'tusks' which are usually absent in the female. The canine teeth are on each side of the jaw some distance behind the incisors, two in the upper jaw and two in the lower. In addition to this, in both sexes there are four small 'wolf' teeth, one in front of each first molar. The two in the upper jaw are usually shallow rooted and may erupt during the first six months, whereas those in the lower jaw seldom erupt at all; precise details are difficult to give as these teeth are retrogressive being leftovers of the prehistoric horse. Those in the upper jaw are usually cast at three years by the erupting permanent molars, and if not they are best removed as they are of no value. It used to be said that their presence affected the horse's temperament and eyesight, causing it to shy but there is no scientific proof of this; in some cases they might interfere with bitting.

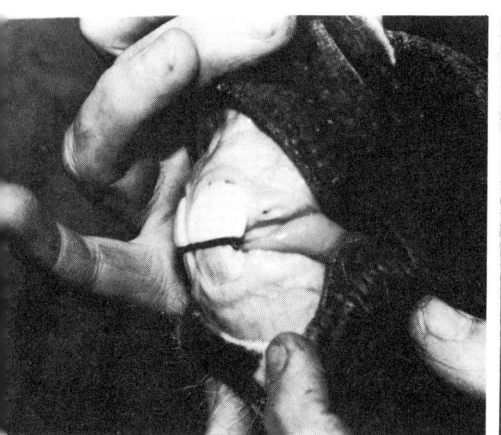

Above left: *A foal's first milk teeth appear at or within a few days of birth. Here the central incisors are well grown at six weeks.*

Above right: *At eight weeks the lateral incisor milk teeth have erupted and are growing.*

The front, incisor teeth are used for ripping or tearing off grass or hay, and the side teeth – the molars – are for grinding up the food. Like humans, the horse is not born with its permanent teeth but first grows temporary or 'milk' teeth which are gradually shed and replaced by permanent teeth at between two and five years of age. A foal will have its first milk teeth at or within a few days of birth. These consist of two central incisors top and bottom and three premolars or cheek teeth top and bottom along each side. All the milk teeth are smaller and whiter than the permanent teeth that replace them. The following table shows how the teeth develop, and although the time of eruption will vary it forms an acceptable guide.

Many foals will try to graze as early as their very first day if out at grass but will have difficulty in reaching down to the ground in order to reach and bite off the swarth. At a week old they will have learned to bend their knees and lower their head in between; in this position they look most peculiar, spider-like and deformed. By the time they are two or three weeks old they will be grazing considerable amounts, and during this time will have learned to drink water placed within reach. From this time on they will graze properly, and if the mare is receiving a small grain ration that is also within reach they will begin to eat this as well.

At birth to 1 week:	2 central incisor milk teeth upper jaw; 2 central incisor milk teeth lower jaw; 3 premolars each side of upper jaw; 3 each side in the lower.
4 to 8 weeks:	2 lateral incisor milk teeth appear in each jaw.
8 to 10 months:	2 corner incisor milk teeth appear in each jaw.
1 year:	All 12 temporary incisors are present and in wear. There are now 4 cheek teeth in each side of the upper and lower jaw – those that were present at birth or soon after plus the first permanent molar top and bottom each side situated at the back of the 3 premolars.
2 years:	The full set of temporary incisor teeth are still present. The second permanent molar is present, making 5 cheek teeth top and bottom each side (3 premolars and 2 permanent molars).
2½ years:	The temporary central incisors are replaced by the permanent central incisors.
3 years:	The permanent central incisors are in wear. The front 2 premolars top and bottom each side are replaced by permanent molars, making 4 permanent molars and only 1 premolar left in the middle of these cheek teeth.
3½ years:	The permanent lateral incisor teeth appear, and by 4 years will be level with the permanent centrals.
4½ years:	The temporary corner incisors are replaced by the permanent corner incisors. The last 2 permanent molars appear – one pushes out the only remaining premolar and the other erupts behind the others, completing the set of 6 permanent cheek teeth along each side top and bottom.
5 years:	The horse's full set of permanent teeth, both incisor and molar, are all up and in wear.

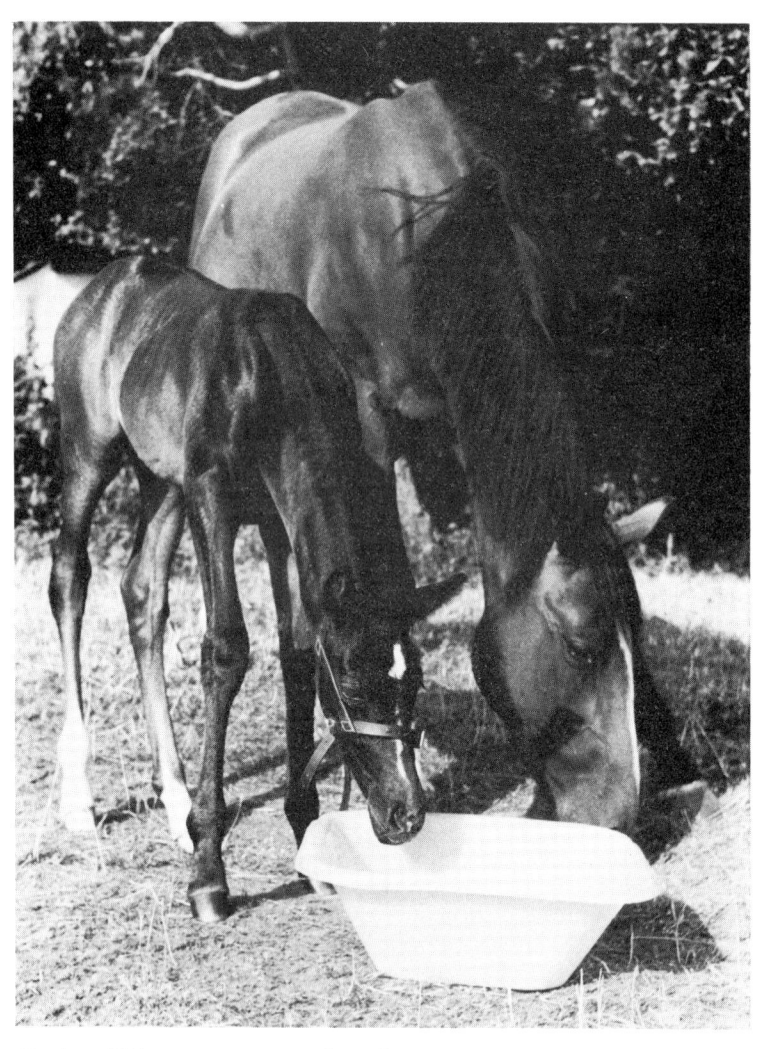

Foals will be grazing considerable amounts by the time they are a few weeks old, and will also begin to eat some of the mare's small grain ration if this is placed within reach.

Yearlings

Yearlings and youngsters up to two years old can have increased handling to teach greater confidence by periods of schooling that are longer than before but still not of a length that bores or tires them. Fit a proper headcollar or use a lungeing cavesson and lunge rein, and also introduce the lungeing whip to the procedure, but only use it to 'gentle' the animal along or signal your intention, and never punish or frighten the youngster with it. Keep the lungeing periods to about ten or fifteen minutes and only about once a week, and don't worry if several weeks or months go by with no lungeing at all as serious training will not come for some time yet. Try to use any periods of lungeing to calm the animal and give it confidence; if it becomes over-excited stop at once and calm it down until it learns to walk without stress and tension. On no account risk injury to its joints at this tender age by letting it cavort about on the lunge, which will exert quite different forces on its limbs from those it experiences when it is free and unconstrained.

Two- and three-year-olds

Between the age of two and three years you will need to break in the youngster, and when you do this will depend on how well developed it is. Big, bossy and bold individuals are better broken nearer two than three years old as they can otherwise become too much of a handful. From weaning the foal should have been taught to stand still and accept being tied up in its box while you move around it asking it to move over. If you have not done so before, it should now be taught to accept a bit fitted to a proper bridle, with a headcollar fitted over this so that the animal can be tied up for short periods. This will enable it to 'mouth' the bit while standing quietly; some people use a plain snaffle while others, of whom I am one, prefer at first to use a solid bit of the type that has rollers or 'keys' in the centre, which encourage the young horse to relax its mouth and produce saliva, thus developing a 'soft' mouth. Make sure it is fitted correctly and that the bridle is adjusted so the horse is comfortable, and place a strong headcollar over the top. Use a strong rope and tie the horse to a solid support that has no risk of breaking should the horse at first struggle and pull back. It is *most* important that the animal is *not* tied up to a movable object; gates are particularly hazardous as a horse can lift them off their hinges and then in frantic efforts to get loose the most horrendous accidents can happen. This cannot be emphasized too strongly, as I know of owners who have had their horses killed or very badly injured from a few seconds' lapse of thought about this. Not only people that are new to horsekeeping are prone to this mistake, as I witnessed it being done by an internationally famous 'trainer' during an instructional television programme.

With patience and common sense you will succeed in getting the young horse accustomed to the bit without being frightened by it, and thereafter lungeing can be carried out with the bit in place. This part of the young horse's training was fully described in book 3 of this series, *Riding and Training*, so I will not repeat it here.

The next thing required is to introduce the saddle, and then

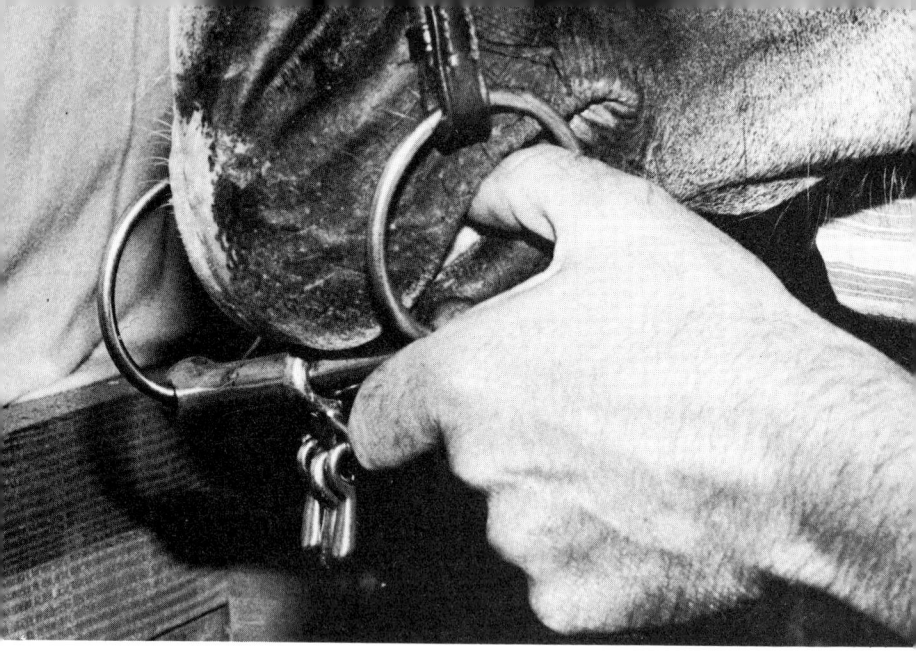

A young horse being bitted for the first time: a plain snaffle with a roller and keys is used for this to encourage the production of saliva.

In this photo the horse has obligingly opened its mouth, showing how the keys fit on the tongue and do not interfere with the teeth when the mouth is closed.

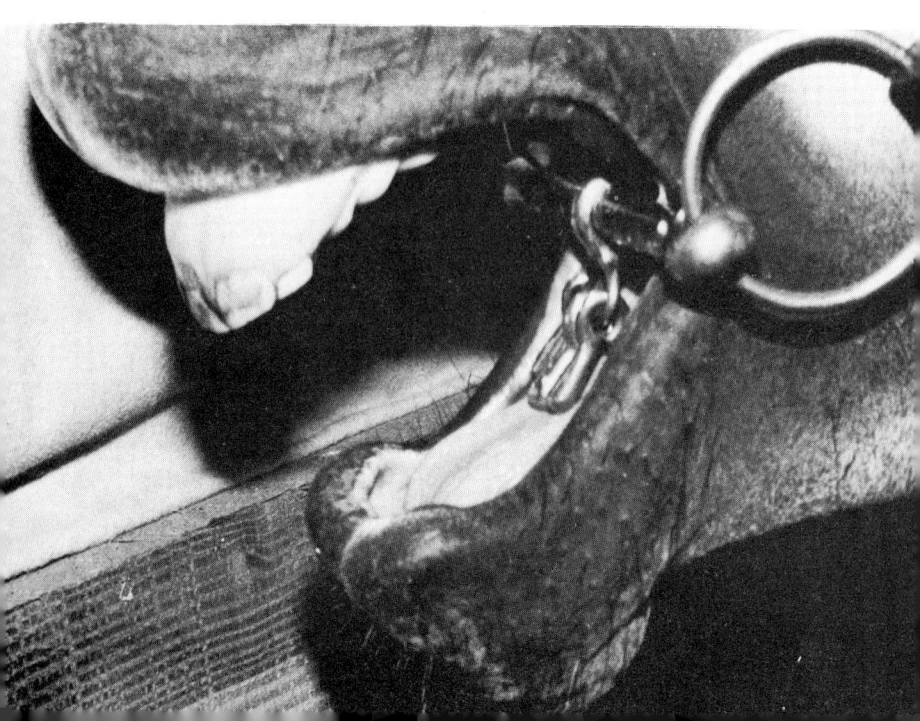

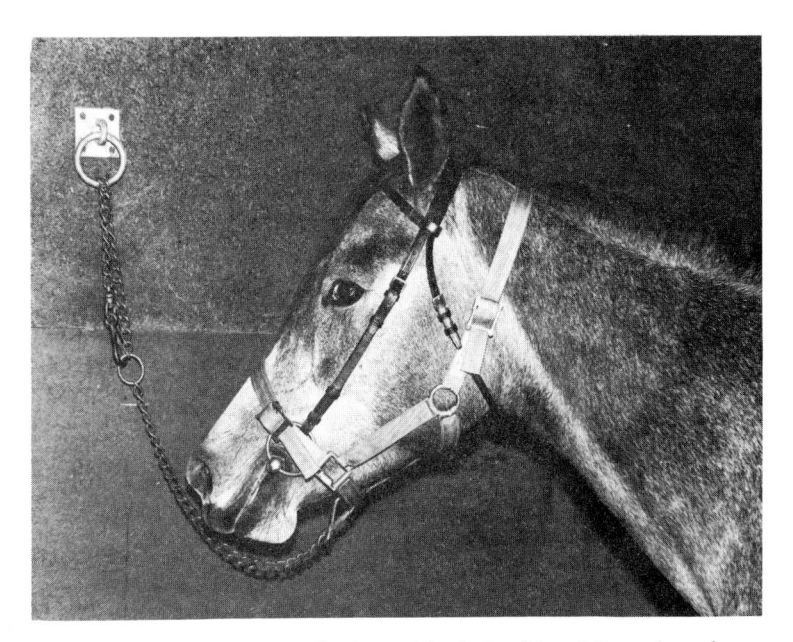

The horse standing quietly 'mouthing' the bit while tethered to a secure fixing with a well-fitting headcollar over the bridle. Note the absence of obstructions, which might cause the animal to get caught up and injure itself.

finally to teach the horse to accept and carry the weight of a rider. If it has been well handled from the outset and is accustomed to the feel of pressure on its back and around its middle as I have described there should be no great difficulty with this last stage of its breaking. Begin by introducing the saddle to the horse at feeding times, when it should be placed on the bottom door or somewhere that the animal can get used to the sight of it and sniff it to assure itself there is no danger. This needs to be done several days before you intend to saddle the horse, and during this period it is a good idea to put a strap or surcingle around the horse when it is tied up inside the box. These actions are carried out gradually and calmly so that each step is accepted and there is nothing to frighten the animal; done in this manner the horse will have no fear of you or the equipment and nine-tenths of the battle will be won before the actual saddling takes place. When you decide the time has come to fit the saddle it is best to have an assistant, and to carry this out somewhere with plenty of room and free from obstruction – perhaps the schooling arena. Tack up the horse in its lungeing cavesson and lunge it for a while to settle it down, then have your assistant take over the lunge rein while you calmly

When introducing the saddle for the first time choose a place with plenty of room, and first lunge the animal to settle it down. Have an assistant take over the lunge rein as you approach the horse and calmly place the saddle on its back.

Lightly tighten the girth, and then walk the horse around to get it used to the feel of the saddle. Stirrups can either be removed or run up as shown.

A bridle is fitted and quiet lungeing continued once the horse accepts the saddle and girth without trouble.

The final stage, with the assistant holding the horse steady as the trainer gradually introduces his weight over its back.

approach the horse and reassuringly place the saddle on its back. As this is accepted slowly girth it up and then carry out a short lungeing programme for a while to settle it down, without the stirrups fitted as you do not want anything flapping about that will frighten the horse at this stage; the girth should be where it causes least discomfort – that is, just behind the elbows and tightened just enough to prevent the saddle from slipping. Carry on like this daily for about a week, during which time the stirrups can be fitted – first in the run up position and then being allowed to dangle. When all of this has been accepted without fuss and the girth has gradually been tightened until it is tight enough for mounting to be done, the time has come to mount up for the very first time. First lunge the horse to settle it and then have your assistant hold it while you mount up. Here different people adopt different techniques; some have a leg up and merely lay across the saddle, while others gently take their weight on the mounting stirrup and remain there to check the horse's reaction before swinging their leg over. Be prepared for the animal to move about, as it will have to adjust to your weight and find its balance. I think it best at this stage not to have the horse bridled with a bit and reins but held quietly in its cavesson and lunge rein, as if it does fidget or buck the rider's first reaction will be to snatch up the reins, and this will make matters worse. If things go smoothly, however, dismount and bridle the horse, then repeat the procedure holding the reins; use a plain snaffle and not one with 'keys' so that it does not fiddle about with its head at this stage. If all goes well, have your assistant lead the horse around with you mounted, and from here on this training can be progressively advanced. Remember, though, that the horse will need to adapt and adjust to the rider's weight. This must be considered with all future work, never allowing the young horse to become tired or strained in any way.

Bitted and broken to saddle by gradual progessive training in the way described over the first two or three years of its life it will become a reliable and pleasurable animal for many years to come. If serious training is not begun until it reaches four or five years of age so that its joints and ligaments have time to mature without undue prolonged strain, it will also remain sound in these areas throughout its life.

7 Feeding

Feeding programmes adopted by individual breeders will differ according to particular breeds, and will vary considerably between, say, a mountain pony breed and the Thoroughbred. I will endeavour to give the readers a sound basis for the feeding required for both mares and young stock in such a way that they will be able to deviate from it to suit their own particular circumstances without the danger of incorrect feeding, which can lead to trouble.

Brood mares

A brood mare out at grass during the summer need not normally be given additional feeding, though this of course depends on how plentiful and lush your grass is. It is not advisable to overfeed during this period, especially with grain – to do so will not help her produce a large, healthy foal but instead may lead to various digestive and other troubles. With very young or old brood mares I like to give a small evening feed of grain throughout the summer of about 3 lb (1.25 kg) consisting of equal parts of crushed barley and rolled oats mixed together with molassine meal and a tablespoon of cod liver oil. The mare gets used to coming for this and it gives an opportunity to look her over at close quarters and handle her. Should she be reluctant to come for this feed I never persist, as it is safe to asume she does not need it. As the autumn comes and winter approaches the grain feed each evening is gradually increased until the bad weather sets in. From then on I have the mare in at night and feed her concentrates night and morning, turning her out during the day unless the weather is really bad with driving wind and rain. The hay rack in her night box is kept full so that she can take in what bulk she requires. I do not use a high protein seed hay for this but instead use my own meadow hay made the previous year. Fed like this and given daily exercise out at grass a brood mare will keep herself fit and in good condition.

Naturally if you intend to ride her during this period or for some part of it your feeding will have to take this into account.

The daily amount of concentrates fed will depend on the size and type of mare, and will probably reach 6–10 lbs (2.5–4.5 kg) during the winter months. This is where your eye will be the best guide: if she gets too fat or over-excited cut down the concentrates; if her condition falls off or she looks generally dejected then increase the ration. For those wishing to use a manufactured cube type for convenience the recommended rates will have to be referred to, but I feed equal portions of crushed barley and rolled oats that I buy in small enough quantities to ensure its freshness. I then add appoximately ½ lb (225 gm) of bran and some molassine meal to the feed night and morning, and also continue to add a tablespoon of cod liver oil alternating with a measure of vitamin supplement as recommended by the manufacturers. During very cold weather I boil four or five handfuls of whole barley and mix this in hot to make a warm meal in the evening. Fed in this way the mare should winter well and be receiving all the nourishment and trace elements she requires to form a strong, healthy foal. As spring approaches and the weather turns warmer the length of time she spends out will increase and she can be gradually cut back until she is out at grass again full time.

It is a mistake to think that as her time approaches a brood mare needs feeding for two. Some mares, however, are poor doers, and older mares especially can be a problem at this time and will go back in condition unless their feeding is very carefully monitered. This is also true if you breed from a very young mare that is only two or three years old when put into foal. These will sometimes do well during their pregnancy but in their first winter after foaling, particularly if the foal is still suckling, they can get quite poor. Don't worry too much about this unless it gets very bad as they will soon pick up the following spring or when the foal has been weaned. However, as they themselves are still growing you must be sure that their diet is not lacking in the vital vitamins and minerals that are essential for their healthy growth, and there can also be a danger of getting anaemic at this time. Salt and mineral licks must be available at all times, as well as a good supply of fresh, clean water. When food supplements are added to the feed they should prove helpful in preventing deficiencies, but it is essential to make sure that they are fresh and not damp or stale. Manufacturers state that these products retain their vitamin levels up to twelve months from the date of manufacture. I personally doubt that they are good for that length of time, and in any case it depends on their being correctly stored in dry, cool conditions. Many retailers leave them on a shelf in a warehouse where they stay through all temperatures and often in damp conditions. It is

Two well-mannered brood mares eat from the same bowl without squabbling: well-disciplined, contented brood mares possessing good temperaments will almost invariably produce healthy foals of similar good nature.

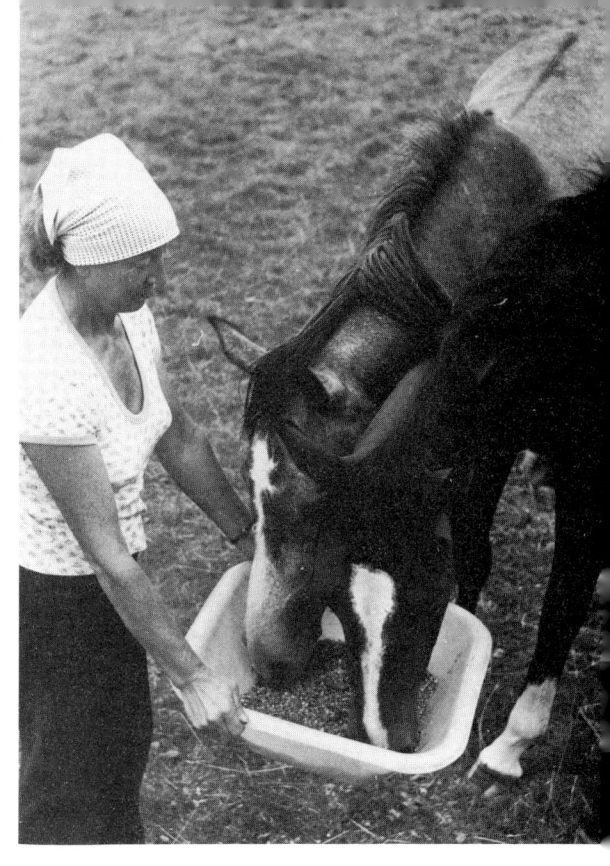

therefore wise to check that they are well within the time limit and buy them from a source that has a quick turnover. Milk pellets are another product often recommended for brood mares and mares and foals as they contain a high protein content together with vitamins and minerals. They are expensive to feed and I have found that a number of mares (and other horses) dislike them and will leave their food when even a small amount, well short of the recommended level, is added. This may be due to their smell, as horses are very sensitive in this respect, or may at times be because the pellets are sour, though manufacturers will never admit it. I have also had cubes refused on occasions because the milk powder used to 'bind' them had gone sour. Even a slight tainting that cannot be detected by the human sense of smell will be enough to put a horse off, and if you feed these products and have them refused it is more likely to be because of this than that the horse is not hungry or is just finicky.

Mares and foals

After foaling, continue to feed the mare as before and if you are giving her a small grain feed while she is out at grass during the spring and summer then feed this in a large container that leaves room for the foal to feed as well. Almost from the beginning the foal will be inquisitive enough to try and eat some and in a very short time it will be enjoying little mouthfuls. If you time this so that it is given in the cool of the evening it will add to their contentment as they can enjoy the food without the bother of pestering flies. These evening feeding sessions will give you much pleasure as you cast your eye over the mare and foal, and will teach the foal to eat concentrates so that after weaning there is no problem with changing to solid food.

As the year reaches equinox and the days start to shorten I begin bringing the mares and foals into an open barn to feed at one long trough where they soon learn to stand in place and discipline each other. In this way the older mares teach the youngsters manners, and provided that there is enough room for all to eat without the need to squabble I have never had any trouble. The hay racks above the manger are always kept full, and mares and foals eat what hay they need. Later on in the year when extra grain is fed to offset bad weather and the lack of grass I revert to bringing each mare and her foal into a separate box where I can monitor their individual requirements. This is necessary so that some foals are not overfed on grain, especially oats, as this can cause epiphysitis – an enlargement of their fetlock joints – and can lead to lymphangitis. A calcium/phosphorous imbalance can also cause this problem, so too much bran in proportion to other grains should also not be fed. When calculating the amount to add to the mare's grain feeds in order to feed both her and the foal much will depend on the size and type you are breeding. However, as a general guide I feed at the rate of about ¾ lb (340 gm) per month of age, so that at six months the foal is getting 4½ lb (2 kg) of mixed grain. The exception to this is a foal born very early in the year, who is therefore still on good summer grass at six months, when the scale of increased grain ration should be delayed so that the 4½ lbs (2 kg) per day is reached in the autumn (fall).

Feeding requirements at weaning – mares

When the mare and foal are separated at weaning time keep the mare out of sight and hearing of the foal for up to a week or so, depending on how she settles. It is safest to confine her to a loose box, which will also enable you to dry off her milk. For the first few days restrict her water, letting her have only a small drink morning, noon and night. It will depend on the size of the mare as to how much she can be given, but don't let her gulp down unlimited amounts. Another method is to leave her some in a bucket in the morning and only replenish it at night so that she is only drinking about half her normal daily intake. Feeding will

If the mare is given an evening grain feed in the field her foal can be tempted away to eat some that is offered separately. This will encourage friendliness and teach it to eat grain so that after weaning there is no problem changing to solid food.

depend on the general condition of the mare, her age and the time of year. If she is in good condition cut out all the concentrates and grain feed she has been receiving and for several days let her have only hay, perhaps coupled with an hour's grazing as she settles enough to be let out. Her normal feeding can gradually be restarted as her udder shrinks and the milk ceases to be produced. As she is put back onto grass, if this is still lush gradually allow more time each day until a normal routine is reached. Most mares will stop producing milk if this routine is followed, and within two weeks her bag will have shrivelled and completely dried up. Do not resort to drawing off milk by milking her by hand unless her udder becomes very swollen and painful. Even then only a little amount to relieve the pressure should be taken off, as the mare will only go on producing more milk if it is regularly being taken away.

Feeding foals after weaning

Kept shut away for a few days, the foal should have plenty of fresh water and hay available. The same daily amount of grain or concentrates that it was receiving prior to its being weaned should

Mares and foals can be taught to eat together from a large manger without squabbling, and disciplined to stand in place, so learning good manners. Note the hayrack above the trough, which is low enough for the foals to reach.

also continue to be fed. If it is six months old the grain or concentrates rations will probably be about 4½ lb (2 kg), and if weaning takes place in the late autumn then as winter approaches the ration can be gradually increased until it is getting approximately 6½ lb (3 kg) of mixed grain together with hay and grass, according to weather conditions. Split the grain feeds into two or preferably three separate feeds, and keep a careful watch for any signs of over-feeding such as hot and swollen lower legs and joints. The amount the foal can safely be given will depend on its size and type, but if in doubt it is better to under feed grain and cubes, etc. slightly, and provided that there is an ample supply of good hay, and grass when this has feed value, it will do well. Any extra supplements that were fed prior to weaning should be continued, and a few handfuls of whole barley that is boiled and added to the feed will be beneficial during the bad weather, as will soaked sugar beet pulp. As spring comes and the foal is given more time out at grass the grain and concentrates can be gradually reduced until it is receiving one small feed in the evening. Don't worry if it goes off its food for a day or two after weaning, as this is a traumatic time and quite often the foal will at first be too upset to eat. This period will soon pass, and any slight setback in its development will be made up later. This is where it is better if two foals can be weaned at the same time and kept together as the company will help to prevent loneliness, and although at the very beginning it can be better to shut them up separately they can be housed together after this. With all but the hardy pony breeds the first winter after weaning should be dealt with by bringing the foal in at night on to a dry bed where it can lie down in comfort; this is a crucial time in its growth and development and although extra feeding might otherwise be given it will not compensate for exposure to continuous bad weather without adequate shelter.

Feeding yearlings

From the age of one year onwards feeding will increase according to size and breed. After their first winter as foals they will be turned out the following spring, and as the days lengthen will remain out full time and if on good grass no extra feed will be necessary. As yearlings they will fend for themselves, and as far as feeding goes management of the grass is all that is required. As they enter their next autumn and winter comes, the extra feed they require will depend on the weather and if they are to be in at night or wintered out. Again it is not possible to be precise about their needs, but generally speaking 7 lb (3 kg) of grain or concentrates can be taken as a guide to daily feeding, with allowances made above and below this for the severity of the weather and the breed. Never feed this all at once as there will be a danger of colic or digestive troubles if more than half is fed at one time; always leave an interval after concentrates are fed before turning out onto wet grass. As much good hay as they can eat should always be available, and of course fresh water is most important. Fed and cared for in this way yearlings will winter well and grow on to become healthy and well developed horses.

When out at grass youngsters will do best in the company of others, but if they are with older horses make sure they are not bullied. Sometimes they can be chased about in more than just playful antics, and this can make them timid or spiteful and will undo all the good of their previous handling; try to make sure they are happy and contented with their companions. Finally, remember to look to worm control as good feeding practice will be largely wasted if you are simply feeding the worms.

8 Stallions

There are no great problems involved in keeping your own stallion provided that certain essentials in their management are attended to. Some breeds will need more special handling than others and individuals will vary within the same breed, but all stallions will possess virility, strength and courage to a higher degree than other horses and often show more intelligence. Most breeds, including Thoroughbreds, will not normally begin their stallion duties until the age of four years, but they can start much later than this and will go on well into their twenties. Before the age of four care must be taken with colts and young entires, as from as early as one year old, and certainly before they are two, they will be capable of serving a mare should they be free to run together.

Many small breeders of non-TB breeds will keep a stallion and train it for use in various pursuits, even using it to ride escort with other horses; some TB stallions are trained in dressage and jumping with no trouble if sensible handling is followed. This solves the exercise problem, as a ridden stallion will be fit and healthy to perform his task during the season and will also benefit from having been used to the company of other horses. If he is not to be ridden he will need to be walked out or lunged unless he is to be turned out with his selected mares. If no exercise is given during the periods he is kept stabled he will need to have access to a small exercise paddock adjoining his loose box so that he can exercise himself to some extent. Make sure, though, that all the fences and door fittings are very strong and never allow other horses to approach near enough for their heads to come together as this will excite the stallion; if it is a mare he will try to get at her, if a gelding he will resent its intrusion. On the other hand, never shut him away where he cannot see other horses or what is going on around him as this will only serve to make him a neurotic animal. It is a question of common sense and the temperament of the particular stallion. I have hunted and hacked out my two Andalusian

Trying: the mare is taken up to the stallion and their heads allowed to touch. This mare is in season and allows the stallion to sniff and murmur to her. Note their friendly expressions and ear positions; mares not in season will flatten their ears and squeal and kick out.

stallions with mares and geldings and my wife and I rode them together for several years without any trouble. I believe that regular riding of the stabled stallion is the best method of management, and if certain rules are followed there should not be any problem. It is usually best to keep the stallion in front of other ridden horses, and if you ride where moorland ponies roam with a stallion then keep a look out for them and do not ride between the loose stallion and his mares. If you decide to turn your stallion out with mares the fences will need to be at least 5 ft 6 in (1.75 m) high, and hedges should also have a strong fence along the inside, though in my experience when there are other horses around the boundaries the stallion will sooner or later try to get to them, and most probably injure himself.

Handling

In the general handling of a stallion you must be firm and gain his confidence and respect, and never ill treat or try to bully him; nor must you be fearful. The stallion will be quick to perceive the slightest nervousness and will try to take advantage of you. Confident handling will usually obtain his obedience without the necessity to resort to punishment, but if he tries to exert himself and is too wilful then a short sharp reminder with a stick will usually restore the balance. Once he has learned that you need to be obeyed, and so long as he does you will treat him with kindness, then you will not have any trouble. Stallions respond to routine much more than other horses and are much better when managed in this way. Most of them will become quite attached to a particular handler and accept this person as being his superior, often to the extent that a kind of bond will grow between handler and animal. Do not, though, mistake this for what it is not, and never put yourself in a position where he can take advantage of you and cause you injury. Remember the old saying 'familiarity breeds contempt', and although with everyday handling you will treat him like any other horse do not get over-confident to the point of carelessness.

Stallions are not usually given to kicking – certainly the Andalusian breed is not – but they can on occasions rear up. This is easily controlled in most cases provided that the handler is calm and confident, and appears much worse to the onlooker than it actually is. I once had a young stallion that did try to kick me, and as this was in his loose box I immediately chased him round and kicked him back. Thereafter he never once tried it again! The point to bear in mind is that the stallion's natural instinct is to be dominant, as in the wild he will constantly have to defend his position as head of the herd or else submit to another, stronger leader. You must convince your stallion that you are that leader, but you cannot do this by strength as he will always get the better of you when it comes to that. Your overwhelming advantage will be your greater intelligence and guile, and using this you should have no difficulty in keeping him in his place.

Another habit they have, even when well handled, is to nip you –

perhaps when being groomed or in impatience if you are slow to feed them. Although this is usually done playfully rather than in a vicious manner and catches your sleeve rather than your skin, there are times when your flesh gets caught between his teeth, and though only momentarily it is nevertheless very painful. I always scold these attempts and give a sharp smack for anything particularly painful. Any punishment that is metered out must be instantaneous so the animal relates it to the event; a smack several seconds later will only serve to confuse and annoy him.

Management and feeding

A stallion used for stud will not need to be fed any differently from other times, nor will his feeding be different from any other horse. Overfeeding will only over-excite him and make him difficult to handle. When stabled he should not be kept in complete isolation, and there is no reason why he cannot be housed next to other horses, even mares, though not when they are in season as this will make him unduly restless. Always be sure, though, that walls and partitions are extremely strong and that he cannot see through or over them, as if he can he will become too interested in the other horse.

A normal healthy stallion, fit from exercise and fed enough but no more than keeps him in condition, will perform his stud duties without difficulty. The number of mares he covers in a season will vary from stud to stud, but I would say that no more than forty or so can be satisfactorily covered. Although some studs with sufficient staff and resources exceed this number with success, the claims made by some for stallions said to have covered as many as 160 mares in one season are to my mind either exaggerated or leave much to be desired. It is not just the covering, which is usually quick and easy enough, it is the preliminary trying and teasing that takes time and trouble. No mare should be forced to accept a stallion unless she is fully in season and ready to breed. When that time comes she will accept the stallion readily enough, but with some mares this period can be very short – perhaps only a day in extreme cases. This is why in my opinion mares should be tried every day as they approach their period of heat so that one can judge exactly the right time for the service and thus ensure a high rate of success.

A mare has two ovaries, which alternately produce an egg (ovum) that is released when ripe for fertilization. This is the period of oestrus, commonly known as being in season or in heat. A mare will normally come into oestrus every twenty-one days during the breeding season – that is, twenty-one days from the beginning of one heat period to the beginning of the next. The normal duration of the heat period will be approximately seven days, and the ripened egg will be in the correct position for fertilization about three days before the end of oestrus. When ejaculated into the mare a stallion's spermatozoa will normally live up to three days inside her genital tract, where the strongest of the millions of sperms will swim up to the top of the oviduct and wait to fertilize an egg. There are, however, considerable variations in this procedure. A mare may have a weak and long drawn-out oestrus, perhaps early on in the season, or may have an intense one in the height of summer that lasts no more than a day or so. A stallion, too, can vary in the length of time the sperms he produces will live after ejaculation. Some very potent stallions will produce sperms that live inside the mare for as long as seven days, while others may have sperms that only live a particularly short life of a day or two. All this shows why I consider it necessary for the stud to have the time to treat each mare with careful attention so that teasing and trying can be carried out thoroughly.

With a drawn-out oestrus the best time for service will probably be from the fourth day, whereas the second day will be best if the oestrus is short and intense. As soon as I am sure the mare is fully in heat I usually have the stallion serve her every day or every other day until I am satisfied that she is going out of season. This is where both the mare and stallion will help the observant handler. A stallion will communicate his knowledge of the mare's readiness when teasing her by the intensity of his reactions, as will the mare when truly ready. I will deal with this in detail later but here I mention it to point out why I dislike the use of hobbles or similar restraints to force a service on a mare, as this often results in it being given too soon. Not only is it useless to serve the mare at the wrong time but it is also dangerous if force of any kind is used in order to achieve this. I have never used these methods and object to their use. There is always a reason for a mare to refuse to come into season or not willingly to accept the stallion, and it is important to understand them. It may be due to no more than nervousness on the part of a maiden mare, especially if she is only two or three years old, and patient teasing and trying to get the timing exactly right will overcome this problem. On the other hand, some mares' refusal can be due to their instinctively knowing that something is wrong, that she should not be bred from because she has defective equipment for doing so. A veterinary examination will often reveal the cause, and many causes can be successfully treated. Sometimes a hormone injection can be given to release the egg into the follicle to travel down to be fertilized. Some owners will want to put a mare that has foaled straight back to the stallion at her first heat (the foaling heat) after giving birth. This is not a good time as the fertility of the mare is low, but it may be – perhaps with an old mare – that the owner thinks she will not come into season again. If the mare does not come into season the vet can induce ovulation, but here again I am not entirely convinced that this is always a good thing, and careful consideration should be given to whether a foal from the mare is important enough to make it worthwhile.

Teasing and trying

Various methods are used by different studs when it comes to teasing and trying mares. Some have special barriers where the mare and stallion are led up on opposite sides for the trying to take place; stallions will be bridled and bitted and sometimes the mares are twitched; as many as four or even five handlers are often advocated. After experimenting in my early years with various methods I soon found it best to keep it as simple as possible, and as I have already stated I dislike force of any kind. I begin by observing the mares out at grass and noting any that show signs of coming into season, such as constantly lifting their tails or showing interest in other horses. I also ride the stallion past the field or up to the paddock gate and see which mares come over to him. Naturally one does not want to get close enough for their heads to come together, for even the best-mannered stallion will become excited when confronted by a mare in this fashion, especially if she is coming into season. This can be time-saving if you have many mares to deal with, but with some mares it is necessary to catch them up daily and take them to the stallion for trying until it is obvious that they are in heat. When I do this I first put a strong leather lungeing cavesson on the stallion and leave him loose in his box with the top door open. I also keep a large piece of heavy carpet underfelt that I fold and lay over the bottom door to prevent the stallion rubbing himself and to absorb any kicks that there might be from the mare I intend trying. Next I catch up the mare, fit a headcollar and rope and lead her around the yard in front of the stallion's box. Watching for signs, I lead her nearer, let their heads touch and wait for the reaction. If the mare throws up her head and squeals and shows no sign of being in season, or if she kicks, I lead her away and turn her out till the next day. If the squeal is not accompanied by a kick or if the kick is only half-hearted it may indicate that she is just starting her heat period; this can also be the case if she offers no objection but at the same time does not show any definite signs. If she lets fly with her back legs when you persist with the trying you can safely leave her for a day or two. You will soon learn to recognize these reactions and be able to reach the correct diagnosis; needless to say it is best to have the mare's hind shoes removed before carrying out this procedure.

If things look promising I lead the mare away some little distance and have an assistant hold her while I put a bandage on

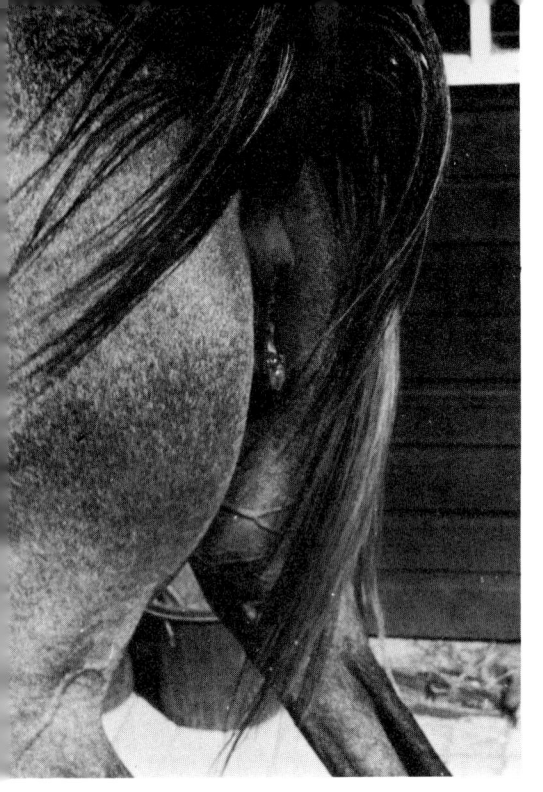

Lifting her tail, she begins 'winking', a sexual reaction of the muscles of the vulva which opens and exposes the inner membranes.

Next she straddles her hind legs and urinates, unmistakable signs that she is ready to accept the stallion.

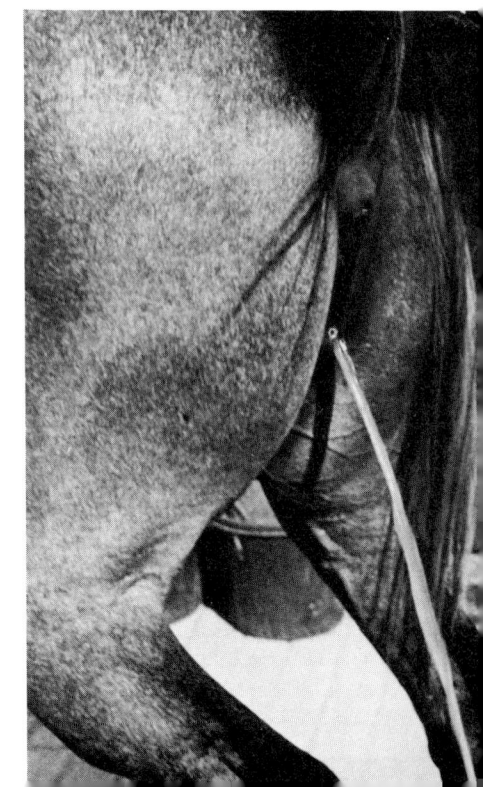

her tail; this will help me to pull her tail away when she is mounted by the stallion, ensuring that no tail hairs are caught up with his penis. I then fetch a sponge and bucket of warm water to which a mild antiseptic has been added, to clean his penis after service. The mare is then led back to the stallion's door, and this time the stallion is allowed to come into closer contact with her and if she reacts favourably he will sniff at her head, then her flanks and finally her tail and vulva; he will probably nibble at her withers and side during this love play. A mare that is properly in season will lift her tail and 'wink', then straddle her legs and usually urinate. Winking is the sexual reaction of the muscles of the vulva, which opens to expose the inner membranes; it is quite unmistakable and is a sure sign that the mare is ready for the stallion. The stallion will now be lifting his head in the characteristic 'flehmen' gesture, an olfactory process in which his lips will expose his teeth, which remain closed, and the top lip curls upwards, compressing his nostrils. This process brings into play accessory olfactory nerves that detect the oestrogen emitted in the mare's urine, allowing him to analyse the scent and confirm that she is in heat; this in turn generates his sexual activity, causing an erection as his sperm production is mobilized.

Without more ado I have the mare led away to another part of the yard within sight of the stallion and then I lead the stallion from his box to mount her. The only form of barrier used is a solid wooden fence or a wall that the mare is led up to, but still allowing her room to move forwards a few steps. The mounting is best not delayed too long as some stallions will swell up and become too large to penetrate the mare: the penetrating end of the penis can occasionally swell to an enormous size, producing a dome some 8 inches (20cm) in diameter; this can be cured by running your hand down it in a series of quick jerks, which will usually cause it to return to normal proportions; there is a danger, however, that the stallion will then ejaculate before reaching the mare.

As the stallion is led up to mount it is sometimes advocated to have him approach from the side or three-quarter angle so that if the mare kicks he can be led away out of danger. I do not favour this, as it invariably means the stallion will mount crookedly and then will have to try to work his way round to find his entry position. This will unbalance the mare and cause a young or nervous mare to become frightened; the stallion will lose his grip and things will have to be started again. In my experience it is far better to have the stallion approach from directly behind the mare, so on mounting she can take his weight centrally to her position and will only need to move forwards a step or two to adjust her balance. If she has been properly teased so there is no doubt about her being in season she will not kick at the mounting stage; if she

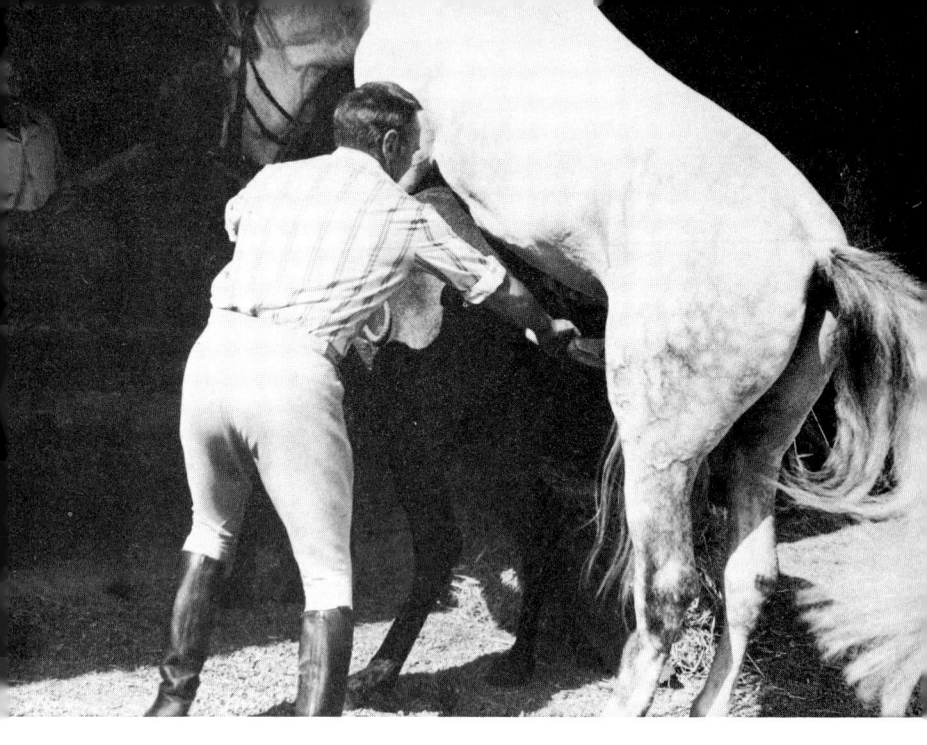

It is sometimes necessary to help the stallion by taking hold of his penis and directing it in the correct position for entry. This is often the case if the stallion is larger than the mare, when he will tend to mount too far up her quarters.

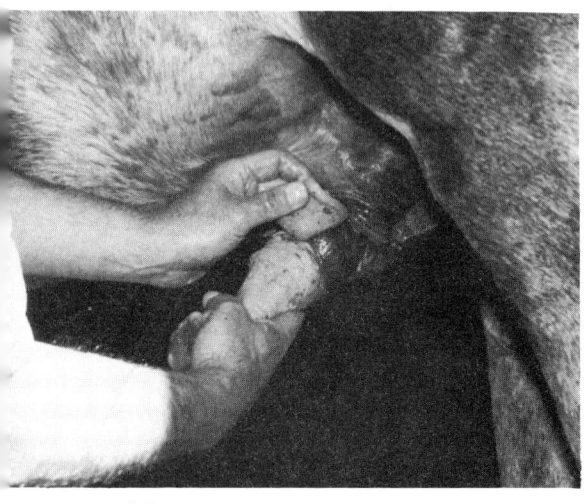

After service the stallion is returned to his stable and his penis and testicles are sponged off with warm water to which a mild disinfectant has been added.

was not quite ready her kicking would have begun at the trying stage before the stallion was led up to mount. Anyone who does not feel quite sure about this should have the assistant hold the mare's head high as the stallion approaches, as this will have the effect of lessening the magnitude of any kick.

Once the stallion has mounted I let go the lead rope to allow him complete freedom of his head and front legs, which are gripping the mare, and I pull the mare's tail to one side to give free access. If the stallion has difficulty I take hold of his penis and direct it into the correct position (with a small mare the stallion may go over her quarters); once inside he will work away, usually 'flagging' his tail up and down. While this is happening I move along towards the front and take the lead rope, and also assist the stallion to maintain his hold by pulling his front leg forwards. It is soon over, and he will go limp and slide off the mare, when she can be led away and he is taken back to his box. The stallion's penis and testicles are sponged off with the warm water and antiseptic, and I also clean his nostrils and the insides of his legs to take away the smell of the mare as he settles down. The mare, meanwhile, is led around for ten minutes so that she cannot stand and strain, when she could eject the spermatozoa. Some loss is inevitable, and this is of no consequence as there will have been millions of sperms released into her, only one of which needs to survive and make contact with the egg awaiting fertilization.

If the mare being covered is much smaller than the stallion it is a good idea to stand her uphill from him as he mounts, and a mare that is taller should be stood downhill. Nothing too steep, of course, but a slight gradient, enough to gain a few inches either way, will greatly assist the stallion to take up the correct position. If the stallion is new to his job do not let him rush the mare but control him at all times; if possible always use a new stallion at first on an older, experienced mare.

All these procedures are accomplished with just two handlers; I have in fact managed alone when putting a sensible stallion to a willing mare, though this is not to be recommended. Two people are best, possibly with a third within earshot in case extra help is needed. More than this tends to lead to people getting in the way, and this is when mishaps happen. A record should be kept of when the mare has been served so that it will be known when her next heat would be due and she can be checked at the twenty-one and forty-two day intervals to see if she has held to the service or returned into season.

9 Ailments

Even when the breeder has taken all the necessary precautions and management is good, mares and foals are still subject to a number of conditions and ailments that are bound to occur from time to time. Once over the actual foaling, and assuming that this has taken place without complications, there are still things to be watched for and properly treated. As explained earlier, it is essential that the foal receives the mare's early milk containing the colostrum that protects it from the harmful effects of infection before it can manufacture its own antibodies. The mare should have been immunized against tetanus and kept in the same environment before the foaling so that her early milk will provide the vital resistance that the foal needs. Within the first few weeks the foal should be given its first tetanus toxoid injection, followed about four weeks later by the booster jab.

Joint ill or navel ill

At foaling time, as soon as the umbilical cord has ruptured the severed end that forms the foal's navel must be treated with an antiseptic disinfectant to prevent infection. Iodine powder is sometimes used for this, but a more effective and convenient method is to use the aerosol 'Violet' spray to cover the wound. This should eliminate the possibility of the germs causing the disease known as joint or navel ill from gaining entry through the navel. As the name implies, it is a disease that causes the foal's joints to become swollen and tense, and it usually first appears between one and three weeks after birth. You may at first think that the foal has knocked itself, as it will be lame on one or more limbs, but it will soon become dull and stop suckling and have a general listlessness and a rise in temperature. Swellings of the hocks, knee and stifle joints will often fluctuate, as will the pain, and may even shift from joint to joint or different legs.

If untreated it can be fatal, and even if the joints appear to recover some animals will show lameness later in life when put into work. Veterinary treatment is needed without delay to bring about successful recovery, but obviously it is better to eliminate the need for this by taking the preventive measures described. Foaling inside, especially on dirty bedding, is said to increase the risk from this disease and is one reason why many people prefer to have the mare foal outside. More important is that the umbilical cord should be allowed to break naturally, when the end will close and congeal immediately, thus preventing access by the disease-carrying bacteria.

Umbilical hernia

This condition occurs as a result of the musculature surrounding the navel area failing to close properly, allowing a portion of the gut to poke through. This may not be visible at birth but will be seen a few weeks later. Sometimes it will appear as a soft, painless swelling that can be pushed back; often it will not enlarge and as the animal grows it will be drawn back inside as the musculature thickens and the skin tightens over the area. In this case by the age of between six and twelve months everything will be normal and no action will have been needed. An elongated navel stump, formed by the breaking of the umbilical cord several inches away from the abdominal wall, will grow back without trouble in the same way.

Occasionally there is a danger from a piece of bowel being imprisoned in the external swelling and actually being strangled. In this case the swelling will be hard and painful and usually warm to the touch; an operation will be necessary to correct the problem. Colts are sometimes also born with a scrotal hernia, when a piece of the gut descends through a canal between it and the abdomen. The scrotum will then be visibly enlarged and may continue to get slightly larger over a period of weeks. Sometimes there is no distress, and by the age of a year or less the condition will have corrected itself. If pinching occurs, causing pain, an operation will be necessary, and in any case special care will be needed when castration is carried out.

Castration

This is not an ailment or disease but an operation carried out to remove the colt's testicles, rendering him unable to service a mare and making him more docile and easily handled. Castration can be carried out at any age from a few months onwards provided that

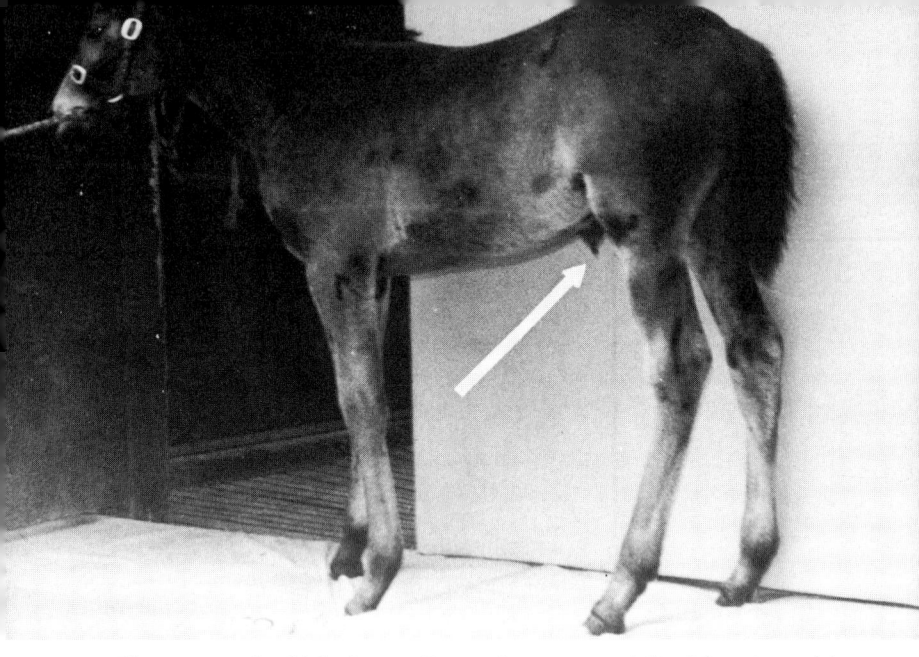

This ten-week old foal is suffering from an umbilical hernia and is waiting to be anaesthetized so that a corrective operation can be performed.

The swelling is hard and painful; the bulge seen here shows where a piece of bowel has become imprisoned outside the abdominal muscular wall. The lower part is the elongated navel stump left from the breaking of the umbilical cord.

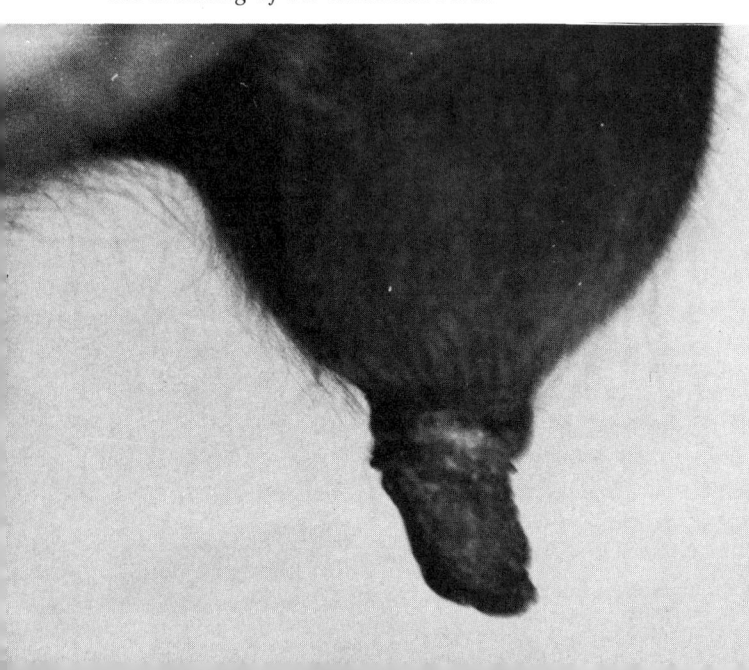

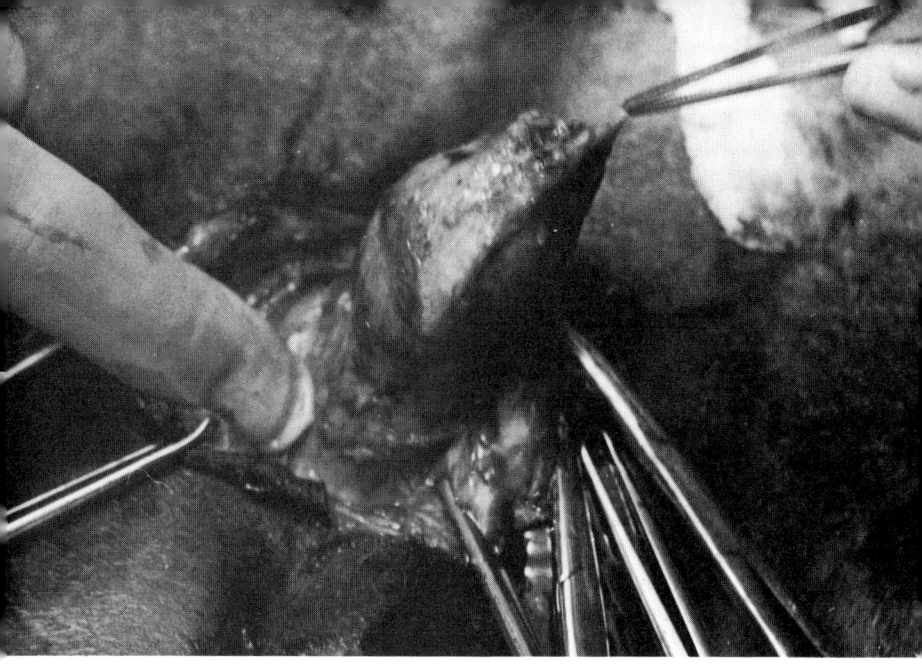

The navel stump has been dissected and the outer skin cleaned away, revealing the breach in the abdominal wall.

The sac of intestinal loop is pushed back into the abdomen; the 'ring' is sewn up and the inside skin tissue is also sewn to give additional support; finally the outside skin is sewn up. The thread used will gradually dissolve away so there is no need for any stitches to be taken out.

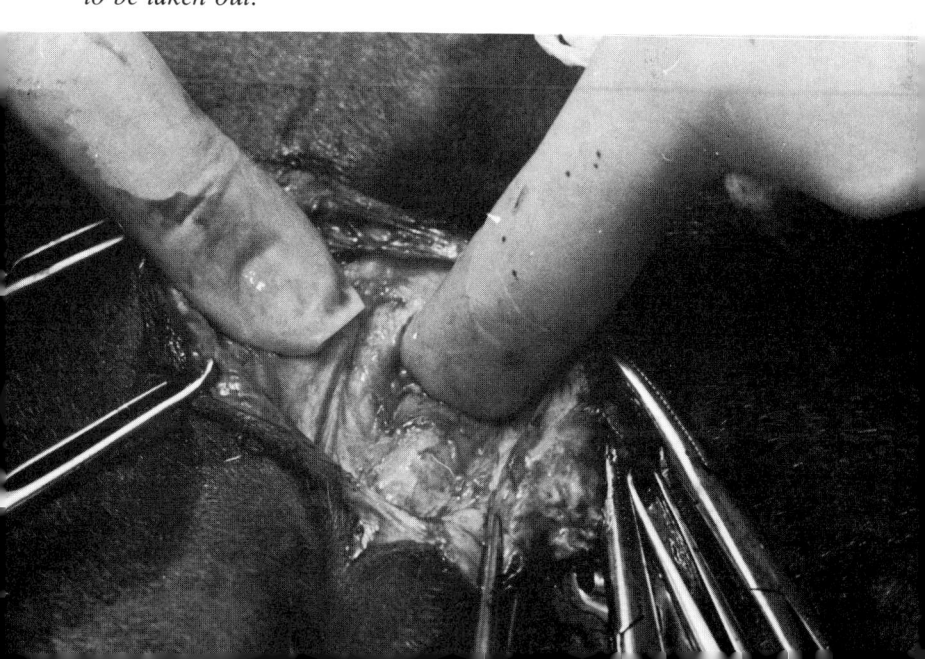

the animal is in good physical condition. Some vets prefer to wait until the colt is a year old, when the testicles are of a size to make the operation easier.

Often testicles are present in the scrotum at birth or arrive there soon after, and when castration is performed early, before puberty, the 'entire' characteristics such as a crested neck are not developed. Some people also maintain that a castrated animal never attains the same richness of coat or full musculature and proud bearing of the entire; for this reason some owners prefer to delay the operation until these characteristics develop. This must be a matter of personal circumstances and choice. Owners not wishing to use their offspring as a future stallion will decide the time to carry out castration by the dictates of management, ease of handling and mixing of young stock; time of year will also matter, as the summer with heat and flies is best avoided. Spring and autumn are the best times to have the operation done, so the animal can be left out to benefit from natural exercise and fresh air which will aid the healing process.

Vets will vary in the method of anaesthetic they use for this operation and many prefer to have the colt taken to their surgery for a general anaesthetic to be given. Where this is inconvenient the operation can be carried out successfully in an open field using the instant 'knockout' drug Immobilon. The intravenous Immobilon will drop a colt within seconds of the injection, and after the operation the animal will be revived just as quickly by an injection of intravenous Revivon. There have been problems with the use of this drug, causing the animal to suffer a bowel prolapse after castration, but such cases are fairly rare. Two vets should be present when this drug is being administered so that should an accident happen and one vet inadvertently get jabbed with the Immobilon the other vet can quickly administer the Revivon; failure to do this would be fatal.

There are occasions when a colt's testicles fail to come down into the scrotum. This testicle retention is usually unilateral, and the retained testicle is often abnormal – it may be small and flabby or greatly enlarged. If it is not removed such animals will be troublesome, being neither stallions nor geldings in the true sense; they are commonly referred to as 'rigs'.

Mastitis

This condition is an inflammation of the udder which may be caused by bacterial infection or the over-production of milk which is retained. With bacteria it may be caused by abrasions to the teats which are then contaminated by flies or dirty bedding. In bad cases the swelling can extend forward from the udder along the lower abdomen, and the mare will show stress signs such as a rise in temperature, higher pulse and respiration rate. She is also likely to go off her food, may have fits of shivering and appear to walk stiffly with the hind legs. Veterinary treatment will be necessary in these cases to prevent the udder function being permanently damaged. Mares so affected will often try to prevent the foal from suckling by biting or kicking them when they try to do so. Mastitis can also occur at foaling time or at weaning, when the udder will become hot and feel hard and swollen to the touch. The mare may or may not find it painful enough to prevent the foal from suckling when it occurs after foaling; when she will not allow the foal to suckle properly this aggravates the situation as the foal will not draw off enough milk. At weaning time the mastitis can take the form of hard lumps in the udder because milk is still being produced. Try to resist drawing off the milk unless the mare is in great pain and keep her from having grass or more than half her normal water ration until things return to normal

Epiphysitis and lymphangitis

Overfeeding or mineral imbalances such as that affecting the calcium/phosphorous ratio can cause these problems, and overfat foals may develop troubles with their limbs, action and wind. Epiphysitis in foals will cause the fetlocks to become rounded and swollen with inflammation above and below the joints and a distinct tendency for the pasterns to become upright. It will appear in the front and hind limbs and all grain feed must be withheld for a period until the condition abates. Caught in time, this will usually take about a week or so but if no improvement is noted then a vet should be consulted and your feed programme discussed. Lymphangitis will need the same treatment; in this case the problem will usually occur in the hind legs below the hocks, and this region will become tense and hot and lameness may or may not be present. Sometimes the swelling can reach well up the legs into the groin, but a watchful owner should recognize the trouble and take remedial action before it gets to that stage.

Contracted tendons – joint and limb deformities

Foals may be born with contracted tendons due, perhaps, to a faulty position while in the uterus, and others may be born with misshapen limbs. Sometimes this will be so severe that the fetlock knuckles over, causing the foal to walk on the front of its foot; at other times the reverse may be the case and the pasterns are completely let down so that the fetlock touches the ground behind the foot. Other abnormalities at birth include rear cannon bones that are banana shaped or hocks that show pronounced crookedness. Hereditary defects can be responsible for these conditions, but in a great many cases they will right themselves, sometimes within a few days; at other times the limbs often straighten as the foal develops and gets stronger. Except in very severe cases surgical correction is not usually necessary, and one can wait long enough to see how the condition rectifies itself. As long as the foal is able to perform its normal functions – being able to stand and suckle, for instance – nothing other than waiting need be done, though strapping or bandaging can sometimes help.

Abnormalities in the joints of young stock, such as inflammation between the lower end of the cannon bones and the fetlock joint in the epiphyseal junction, will usually be as a result of incorrect feeding. It may be due to a vitamin deficiency but is more likely to be due to an imbalance of calcium and phosphorous. This can be caused by feeding excessive grain or the grazing of pastures where artificial fertilization has produced an imbalance in the soil. Too large a concentration of phosphorous will interfere with the absorption of calcium, causing problems at this time of maximum growth in the young animal.

Calcium and phosphorous are two essential minerals in bone formation, but an excessive imbalance, which may also be associated with a vitamin D deficiency, will lead to bone deformity as is seen in rickets. Vitamin and mineral supplements, therefore, should not be haphazardly mixed together as imbalance may result. Phosphorous is abundant in oats and barley and especially in bran; large quantities of bran fed alone will restrict the use the animal is able to make of calcium, iron and zinc, and the imbalance will produce the same result as a straight deficiency of these minerals. The correct ratio is put at between 1.7 to 1 calcium to phosphorous for brood mares and young stock, down to 1.3 to 1 calcium to phosphorous for mature horses in work. The average owner need not worry too much about these strict analytic requirements provided that he or she adopts a sensible feeding programme as detailed in this and earlier books of the series. Do not be misled into believing that because small amounts of certain feedstuffs or supplements can do good, feeding twice as much will do twice as much good. Animals are no exception to the rule that 'too much of a good thing' can prove harmful.

Constipation

Within hours of its birth a foal will need to pass the waste products of its system that were stored in its gut while it was still in the uterus. Foals that are seen to strain constantly without producing dung, and show other symptoms during the first few hours such as restlessness and increasing discomfort due to abdominal pain, sometimes accompanied by rolling on the ground, will need prompt veterinary attention. Liquid paraffin, enemas and the removal from the rectum and pelvic opening of the impacted faeces will be called for. If untreated, severe cases of this constipation in the early stage of a foal's life can rapidly prove fatal.

Constipation at other times in both foals and brood mares can be caused by unbalanced feeding, especially too much concentrates and not enough fibre. The remedy here is obvious, and after readjusting the feed programme a period at grass will prove beneficial as this will provide a natural laxative. Do not, however, turn any horse out on to wet grass immediately after feeding concentrates as this can cause compaction of the concentrated feed and a fermenting of the greenstuff building up behind it, which will cause severe colic.

Diarrhoea – scouring

Diarrhoea or scouring can occur in foals for several reasons: the mare's first oestrus after foaling; unbalanced feeding of the mare or change of location and pasture; infection from worms or bacteria; inclement weather.

About a week to twelve days after foaling a mare will have her first oestrus, and it is not uncommon for this to produce scouring both in her and in the foal. It is due to changes in the milk at this time; as long as the foal remains bright and continues to suckle the phase will pass fairly quickly without special treatment. I have found that foals born later in the year, i.e. July and August, are often unaffected by scouring, and it may be because the grass is less lush than in springtime and the change in the mare's milk is less pronounced and more gradual. At weaning time some foals will have a short period of diarrhoea, perhaps due to the distress of separation or the changeover of feeding; a change of pasture or a prolonged spell of unseasonal weather can also be contributing factors. Again, if the foal remains in good health generally and is continuing to feed well there is usually no cause for alarm or special remedies.

Worm infestation can cause diarrhoea, and there is also the danger of poisonous plants being eaten by inquisitive youngsters. Prevention of this will be brought about by proper management, worm control measures, proper feeding and attention to hygiene, etc., as has been described. Where this is not at fault any cases that do occur will usually be less severe and short lasting. If, however, diarrhoea is accompanied by symptoms such as the foal being distressed, off its feed and with a rise in temperature, then the vet should be consulted as speedy treatment will be needed.

Index

Afterbirth 48, 50
Allantois 46, 48, 50
Amnion 46, 50
Anaemia 82
Ancestry 11–13
Antibodies 50

Barren mares 36
Bitting 71, 75–7
   first stage 62
'Bone' 11, 12
Breaking in 75–80
Breeding stock,
   choice of 12–15
Brood mares, care of 41, 42
   feeding of 40, 41, 81–4
   preparation of 37, 38
   selection of 14, 15, 34–6
Buildings required 29

Caesarean operation 44, 45, 47
Calcium 84, 103, 104
Cannon bone 11, 12, 14, 103
Castration 71, 99, 102
Colic 87, 105
Colostrum 43, 50, 54, 98
Colts 70, 71, 88
Conformation 11–15
Constipation 104, 105

Deformities 103
Diarrhoea 105

Difficult breeders 34
Dung, first (meconium) 51, 54

Epiphysitis 84, 103, 104

Feed supplements 82, 83, 86, 103, 104
Feeding bottle 43, 50
Feeding, yearlings 87
Feet, care of 66
Fencing 25, 27
Fertility 55
Fertilization 92, 97
Fertilisers 23, 104
Field shelters 26–8
'Flehmen' 95
Flies, problem of 35, 102
Foaling box, construction of 29–32
Foaling, haemorrhaging 49, 55
   heat 55
   preparations for 43, 44
   unaided 29, 33
Foals, constipation of 54, 104, 105
   feeding of 84–6, 103–5
   feet 63–5
   first lessons 55–62
   grooming of 62
   newborn 49–51, 55, 72
   rejection of 52, 54
   urinating 51, 54

winter care 67, 68
Foetus 41
Foster mother 43, 52

Genetics 13, 33
Grassland, management of 19, 20, 23, 104

Harrowing 20, 21
Haymaking 22, 23
Hayracks 26, 32
Headcollar (foal slip) 55, 56
Hernia 49, 99–101
Hocks 14, 99, 103

Joints (joint ill) 98, 99, 103

Lactation 51
Laminitis 37
Lead rein 59–61
Lime, the use of 20, 23
Lungeing, introduction to 60, 62
   of young stock 74, 75, 77–80
Lymphangitis 84, 103

Maiden mares 34
Mangers 30, 31
Mastitis 102, 103
Mating season 34, 35
Milk, production of 35
Minerals 65, 82, 83, 103, 104
Mowing 20, 21

Naval ill 98, 99
Naval stump 49, 50, 99

Oestrus 55, 91, 92, 105
Ovaries 91
Over-crowding 17
Ovulation 91–3

Paddocks, drainage of 18
   number of 17, 19
   types of 18
Parturition 44–50

Phosphorous 84, 103, 104
Placenta 47, 48, 50
Problem mares 52, 103

Registered stock 13
Rickets 104
'Rigs' 102
Rugs, the use of 28

Salt 82
Salt licks 65
Scouring 105
Septicaemia 50
Service, timing of the 34, 35
Soil, types of 18, 20
Spermatozoa 92, 95, 97
Stallions, choice of 33, 34, 36
   exercise of 88–90
   feeding of 91
   handling of 90, 91, 95–7
   hygiene 95, 96
   management of 88–91
Stud fees 36
Suckling 51, 56

Teasing 89, 91–5
Teats 44, 45, 51
Teeth 71–3
Temperament 15
Tendons 12
   contracted 103
Testicles 71, 102
Tests for pregnancy 39, 40
Tetanus 63, 98
Trying 89, 91–5
Twitching 52, 53, 93

Udder 51, 52, 70, 85, 102, 103
Umbilical cord 46, 47, 49, 50, 98
Uterus 45, 50

'Verve' 14
Vitamins 82, 83, 103, 104
Vulva 94, 95

Water supplies 23–5, 30–2
Waxing 44
Weaning 69, 70, 84–6, 103, 105
'Winking' 94, 95
Wintering out 27, 28

Worm control 20
Worming 37
Worms 64–7, 87, 105

Yearlings 74, 87